Applied Algebra
and
Trigonometry

Second Edition

Joe Witkowski

Eileen Phillips

Keene State College

Kendall Hunt
publishing company

Cover image © Shutterstock, Inc.

www.kendallhunt.com
Send all inquiries to:
4050 Westmark Drive
Dubuque, IA 52004-1840

Table of Contents

__To the Student__

Welcome to <u>Applied Algebra and Trigonometry</u>! We hope you enjoy learning the material in this text. The material and approach have evolved over many years.

Our motivation in writing this text is to provide you with algebra and trigonometry skills and concepts which will enable you to be successful in your majors of chemistry, physical education, technology studies, applied computer science, physics, economics, management, etc. We also hope the information provided in this text will help you in your life as a contributing member of your community.

The topics in this text do not necessarily follow the traditional path of most algebra and trigonometry texts. We have tried to provide you with information you will need in your non-math majors and in your general education science courses. There are many examples and activities in the text that have been obtained from other disciplines to illustrate the connections between mathematics and these disciplines.

Our goal is that you are able to read the material on your own and understand the concepts presented. We have focused on the concepts and how to understand them, rather than on the technical definitions and cumbersome terminology.

The text is broken down into 15 Units. Each Unit contains several sections. Each section contains information about specific topics, examples on how to solve problems related to those topics, and then activities and exercises are given for you to do some problem solving. Our hope is that this format will help solidify your understanding of the material.

As you go through the text, read explanations critically, making sure you understand what is being said. You should also work out the steps given in the examples. This will help you understand the material better and make the activities and homework easier for you to complete. The solutions to the activities can be found in the appendix. If you have questions, be sure to ask your instructor for clarification.

Good luck and we hope you enjoy your journey through <u>Applied Algebra and Trigonometry</u>.

UNIT 1
The Power of Algebra

<u>**Section 1-1: Why Study Algebra?**</u>

Reason #1 to study algebra:

    Algebra is the language of generalization.

We are often presented with problems where we are asked to identify a pattern that is generated by a situation and then give a general rule for what we see.

Example 1-1.1: How many matches are needed to make 10 squares in a row, the side of each square being the length of a match, as in the following sequence?

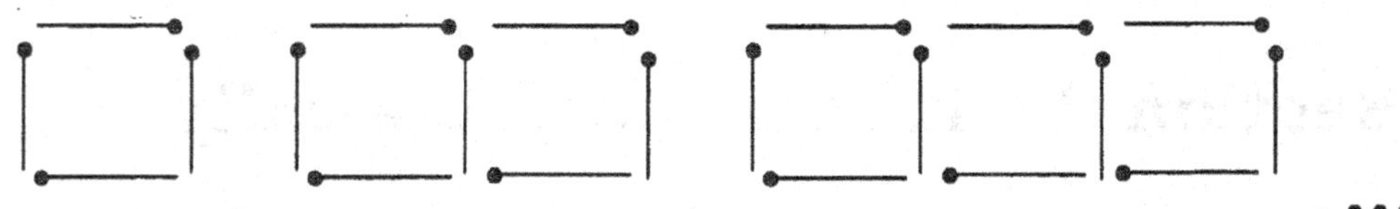

Solution: The most common way to approach this problem is to count the number of matches in the sequence. The numbers in the sequence would be

$$4,\ 7,\ 10,\ 13\ldots$$

We want to find the tenth term in this sequence, which would be 31. It would be useful if we could generalize this problem, so that we could easily find how many matches would be needed if we needed to make 150 or even 958 squares.

In order to generalize the problem we have to understand the structure of the match configuration. One way to see the structure is to redraw it as follows:

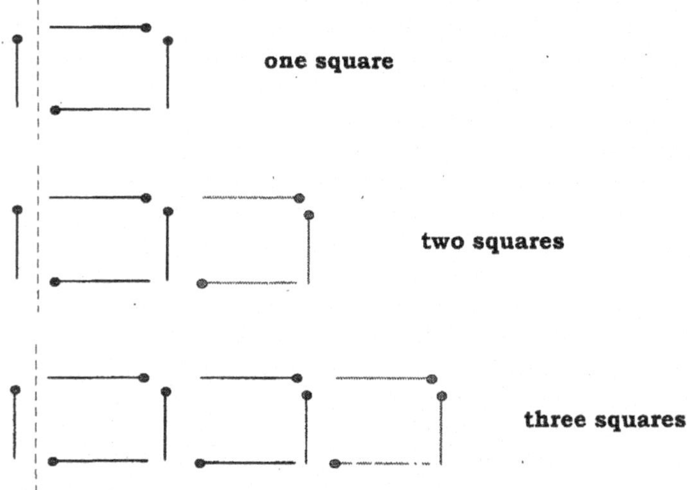

We can see that the first figure requires one match to start with and then one group of three matches. The second figure requires one match to start with and then two groups of three matches. Similarly, the third figure requires one match to start with and then three groups of three matches. We could write this as $1 + 3$, $1 + 3 + 3$, and $1 + 3 + 3 + 3$. Writing more compactly we get $1 + 3 \cdot 1$, $1 + 3 \cdot 2$, and $1 + 3 \cdot 3$. So once we understand the structure we can write the number of matches required for 10 squares would be $1 + 3 \cdot 10 = 31$.

More generally, the number of matches needed for n squares in a row would be *1 + 3n*.

There are other ways to view the structure in this problem. For example, we could count the number of vertical and horizontal matches. Examine the following:

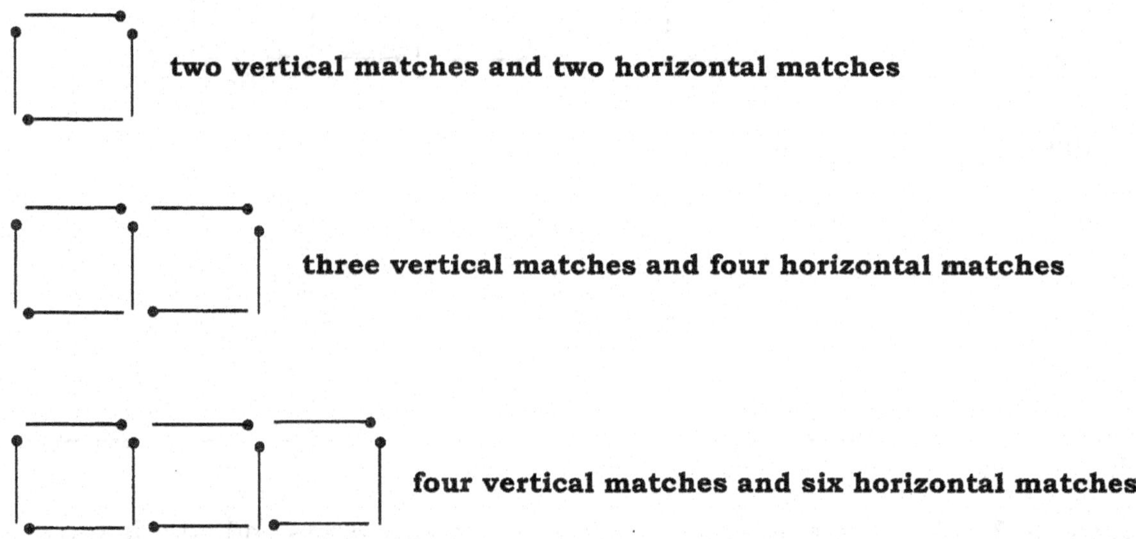

We can generalize this pattern by noticing that if there are n squares in the row, there will be (n + 1) vertical matches and 2n horizontal matches for a total of (n + 1) + 2n matches.

Notice that this formula is different from 3n + 1, but they are equivalent in that they give the same value for a specific value of n (the number of squares in a row).

♦

Activity 1.1

How many matches are required to make N^2 unit squares in a square array as in the following sequence?

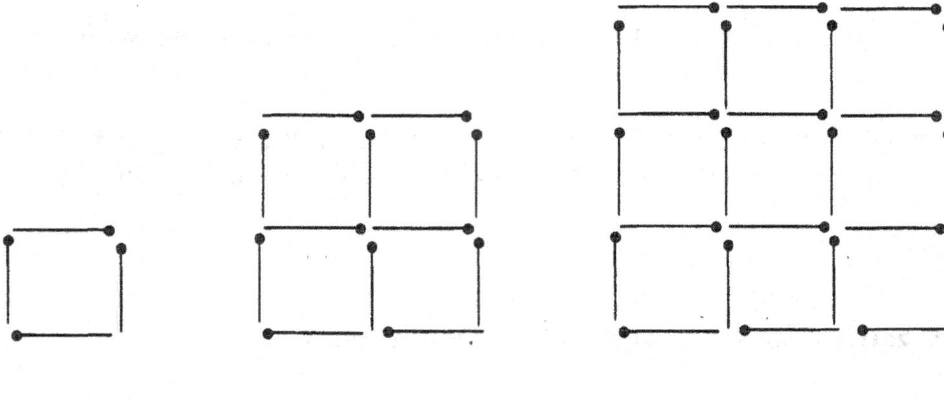

4 matches   12 matches   24 matches

Example 1-1.2: Take a 2-digit number, reverse the digits and add. Is there any pattern?

Solution: First let's try some examples.

Take 27, reverse the digits to get 72, now add.

$$27 + 72 = 99$$

Try a few more examples, and it still may not be obvious what the pattern is.

Let's generalize the problem, in other words, let ab be any two-digit number, a and b are not multiplied, they are each representing one digit in the two digit number. Reverse the digits to get ba and then add.

$$ab + ba = 10a + b + 10b + a = 10(a + b) + (a + b) = 11(a + b)$$

This tells us that if you take any two-digit number, ab, reverse the digits and add you get $11(a + b)$.   ◆

Reason #2 to study algebra:

Algebra enables a person to make general calculations in an automatic way that greatly simplifies large numbers of particular arithmetic calculations.

Suppose you put money in a bank to earn interest. When interest earned is added to the principal at stated periods, the interest is said to be *compounded*.

Compound Interest Formula

$$A = P\left[1 + \frac{r}{n}\right]^{nt}$$

where **A** = **amount present after t years** (principal plus interest)

P = **the principal** (the amount you start with)

r = **interest rate** (must be a decimal, 5% = .05)

n = **number of times a year the money is compounded** (annually, n = 1; semiannually, n = 2; quarterly, n = 4; monthly, n = 12; daily, n = 365)

t = **number of years the money is invested**

Example 1-1.3: Use the formula above to find the amount of money you would have after 5 years, if you initially invest $1000 at an interest rate of 5%, compounded monthly.

Solution: We will be using Excel in the course, so let's use it to solve this problem. If you entered the following, the amount after 5 years would be $1,283.36. Once the spreadsheet is set up, you can change any of the values for P, r, n, or t to find the amount.

	A	B
1	P =	1000
2	r =	0.05
3	n =	12
4	t =	5
5		
6	A =	=B1*(1+B2/B3)^(B3*B4)

Activity 1.2

The formula below gives a method to find the day of the week some event occurred given the date.

$$W = d + 2m + \left\lfloor \frac{3(m+1)}{5} \right\rfloor + y + \left\lfloor \frac{y}{4} \right\rfloor - \left\lfloor \frac{y}{100} \right\rfloor + \left\lfloor \frac{y}{400} \right\rfloor + 2,$$

where

> d = the day of the month of the given date;
>
> m = the number of the month of the year, with January and February regarded as the 13th and 14th months of the previous year. The other months are numbered 3 to 12 as usual; and
>
> y = the year.

The symbol $\lfloor \ \rfloor$ means to round the number inside to the greatest integer less than the number. For example $\lfloor 3.6 \rfloor = 3$. Once W is computed, divide W by 7 and the remainder is the day of the week with Saturday = 0, Sunday = 1 and so on, until Friday = 6.

Use the formula to determine what the day of the week today is.

Reason #3 to study algebra:

Algebra is the language of relationships among quantities.

"A neat example of the value of the machinery of algebra is furnished by one of the classic stories of the history of mathematics. It concerns the exploits of the greatest mathematician of antiquity, Archimedes. Archimedes lived in the independent Greek city-state of Syracuse, which was in Sicily. King Hiero of Syracuse had ordered a crown made of gold but suspected, when the crown was delivered, that the interior was filled with the base metal silver. He asked Archimedes to determine, without destroying the crown, whether the crown was made of pure gold. As Archimedes pondered the problem while taking a bath, he suddenly discovered the physical principle that would enable him to solve it. One historical account states that he was so excited by his discovery that he ran outside naked shouting *"Eureka!"* (I have found it).

As he sat down in his bathtub Archimedes noted that the amount of water that overflowed was about equal to the amount that his body displaced. The fact that a body immersed in water displaces water equal in volume to the space the body occupies is of course obvious. But just such a casual observation often suggests deeper and more useful thoughts to minds that are active on a problem. Minds in this state seem to take advantage of every idea that occurs. In this case Archimedes saw how he could use the observation he made.

Bodies of equal weight are not necessarily of equal volume, for example, a pound of silver occupies more volume than a pound of gold. Hence, Archimedes reasoned, a pound of silver would displace more water than a pound of gold. He formed two masses, one entirely of silver and one entirely of gold, and both had the same weight as the crown. He then put each into a pail of water filled to the brim. In each case the water overflowed but, as expected, far more water overflowed when the mass of silver was immersed. On immersing the crown in the filled pail, Archimedes found that it displaced less water that the silver and more than the gold. By this simple quantitative observation Archimedes knew that the crown was some mixture of gold and silver." (Kline, page 63)

Example 1-1.4: Archimedes wanted to know how much silver and how much gold were in the crown. Suppose the crown weighed 10 pounds, and suppose 10 pounds of silver displaced 30 cubic inches of water and 10 pounds of gold displaced 15 cubic inches of water. The volume of water that the crown displaced was 20 cubic inches. How much gold and silver were in the crown?

Solution: We can use algebra to write the relationships between the two variables and then solve the equations.

Let w_s be the number of pounds of silver in the crown and w_g the number of pounds of gold in the crown. We are given that

$$w_s + w_g = 10.$$

In addition, w_s pounds of silver will displace $(w_s/10) \cdot 30$ or $3w_s$ cubic inches of water. Since 10 pounds of gold displaced 15 cubic inches of water, w_g pounds of gold would displace $(w_g/10) \cdot 15$ or $\frac{3}{2}w_g$ cubic inches of water. Since the crown displaced 20 cubic inches of water, we can write,

$$3w_s + \frac{3}{2}w_g = 20.$$

Solving these equations we get

$$w_s = 3\frac{1}{3} \text{ pounds and } w_g = 6\frac{2}{3} \text{ pounds.}$$

♦

Reason #4 to study algebra:

Algebra allows you to see the general through the particular.

Example 1-1.5: In a discount store, you see an item advertised at 20% off the listed price, but there is also a tax of 12% to be added on. Which would you prefer, to have the discount taken off first before the tax is added on, or to have the tax added on and then the discount taken off the total?

Solution: You may try a particular example first. Suppose the item cost $100, and you take the discount before the tax. Then the item would cost $100 - 0.2($100) = $80. We would then have to add on a 12% tax, making a total cost of $80 + 0.12($80) = $89.60.

Now if we add on the tax first, the price would be $100 + 0.12($100) = $112. Now we would take the 20% discount to get $112 -0.2($112) = $112-$22.40 = $89.60.

$\blacklozenge$

After looking at this particular example, you may make a conjecture that for the customer, it does not matter; either order gives the same result. Algebra allows us to see why the order does not matter. Let's completely generalize the problem.

Suppose that the price of the item is P, the discount written as a decimal is x, and the tax written as a decimal is y. Then we have two cases:

Case 1: Take the discount before the tax

$$(P-xP)+y(P-xP)=(P-xP)(1+y)=P(1-x)(1+y) \quad \text{(distributive property)}$$

Case 2: Add the tax then take the discount

$$(P+yP)-x(P+yP)=(P+yP)(1-x)=P(1+y)(1-x)$$

You can see that the two expressions, $P(1-x)(1+y)$ and $P(1+y)(1-x)$, will give the same result.

By looking at a particular example, using algebra, we can see the general case, where P, x, and y can be any numbers.

<u>**_Section 1-2: Thinking Algebraically_**</u>

It is important that before we proceed you see an example of how to put the ideas of algebraic thinking to work to solve a problem.

Suppose you have an 87 average after 4 exams. What score do you need on the fifth exam to average 90 on all five exams?

There are a couple of ways you may attempt to solve this problem. Some may solve it algebraically by setting up and solving the following equation, where x is the score on your fifth exam.

$$\frac{87 \cdot 4 + x}{5} = 90$$

Others may try solving the problem using arithmetic and would work as follows:

> To average 90 points on 5 exams means you would have to have a total of 450 points. You have $87 \cdot 4$ or 348 points, so you need an additional 102 points.

You can see that we really don't need algebra to solve the problem; we can solve it by just using arithmetic. We might be interested in the following questions:

> What is the highest average that you can attain?
> What is the lowest average that you can attain?

These questions lead us to looking at the relationship between your score on the fifth exam and your average. Let A equal your average for all five exams, then

$$A = \frac{87 \cdot 4 + x}{5} = \frac{1}{5}x + 69.6.$$

We will learn in a later Unit that this represents a linear function with a slope of $\frac{1}{5}$. We will learn how to interpret the slope. In this example the slope tells us that every five additional points you score on any of the tests contributes 1 point to your average. By solving the equation for x we can determine what you need to obtain any average, not just the single average of 90 presented in the original problem.

We will also represent the problem with a picture which can add new insight to the problem. A graph of $A = \frac{87 \cdot 4 + x}{5}$ is given on the next page.

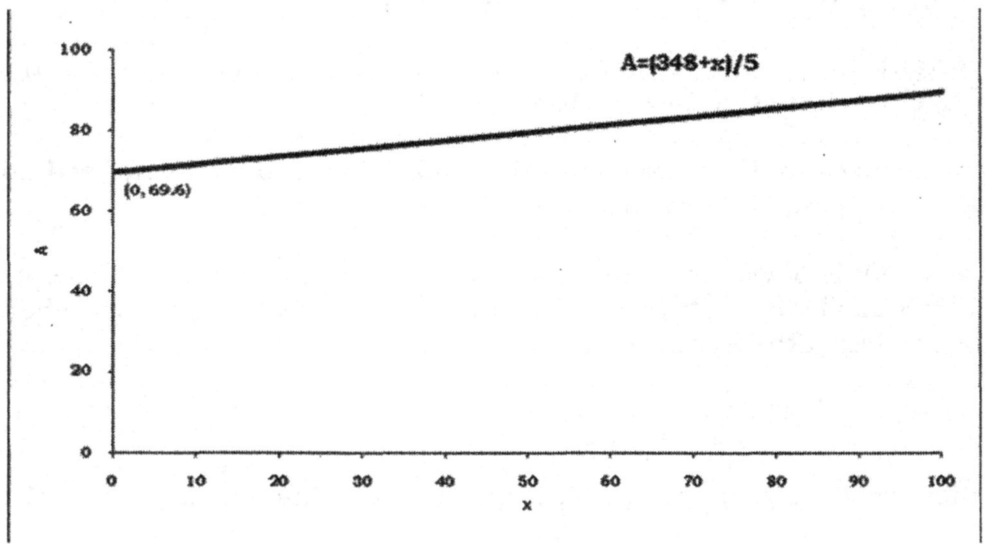

We can see that your lowest possible average is 69.6 and the highest possible average is 89.6. Algebra allows us to solve an entire set of problems at once.

Section 2-1: Introduction

In this section we will cover the critical foundations of algebra. Some stem from topics that you previously covered when you first learned about operations with whole numbers. Things like the associative and commutative properties for addition and multiplication, the distributive property, the order of operations, and understanding fractions.

Some of these properties are so obviously true that you don't realize you are using them. For example, we know that $5 + 6 = 6 + 5$, $3 \times 4 = 4 \times 3$, $\dfrac{2}{3} \times \dfrac{4}{5} = \dfrac{4}{5} \times \dfrac{2}{3}$, and $(-4) \times \dfrac{12}{7} = \dfrac{12}{7} \times (-4)$. Of course it is correct to interchange the order of the terms in addition and multiplication not just for the specific numbers used here. In the next section we will state in general terms these and other properties, which are useful when doing algebra.

Section 2-2: Properties of Numbers

> ***Commutative Property of Addition and Multiplication***
> For any two numbers a and b, $a + b = b + a$ and $a \times b = b \times a$.
>
> ***Associative Property of Addition and Multiplication***
> For any three numbers a, b, and c, $(a + b) + c = a + (b + c)$ and $a \times (b \times c) = (a \times b) \times c$.
>
> ***Distributive Property***
> For any three numbers a, b, and c, $a \times (b + c) = a \times b + a \times c$.

Let's see how we can use these properties of numbers to calculate. The "algebraic thinking" that we will use in these examples will be used throughout the course.

Example 2-2.1: Add 79999 + 269 mentally, stating the properties used.

Solution:
$$
\begin{aligned}
79999 + 269 &= 79999 + (1 + 268) \\
&= (79999 + 1) + 268 \quad \text{(associative property of addition)} \\
&= 80000 + 268 \\
&= 80268 \qquad \blacklozenge
\end{aligned}
$$

In the above example the idea of rewriting 269 as 1 + 268 is an important process in doing algebra. Essentially you are "undoing" something. For example, we know that $3x + 2x = 5x$. Sometimes it is important to "undo" this process, in other words, if you were given 5x, you could write it as $2x + 3x$ or $4x + x$.

Activity 2.1

Mentally compute the following, stating the property you used.

a)   Add 453 + 16999

b)   Add 799 + 1457 + 101

Activity 2.2

Compute the following, stating the property you used.

a)   Compute $17 \times 25 \times 4$

b)   Compute $125 \bullet 37 \bullet 8$

c)   Compute $1001 \bullet 20$

d)   Compute $1001 \bullet 102$

Example 2-2.2:   Use the distributive property to multiply the following:

$$(a + b)(m + n)$$

Solution: Since $(a + b)$ is a number, the distributive property says that

$$(a + b)(m + n) = (a + b)m + (a + b)n$$

Using the distributive property again,

$$(a + b)m + (a + b)n = am + bm + an + bn.$$ ♦

"Undoing" the distributive property is called **factoring**. You can also think of factoring as the process of writing something as a product. For example, when you write 6 as 3×2 you are factoring 6.

Example 2-2.3:

$$3x + 18 = 3(x + 6)$$

$$\uparrow \qquad \uparrow$$

original   factored

expression   form

Notice that factoring is really the reverse of the distributive property.

a(x + y) = ax + ay	*Distributive property*
ax + ay = a(x + y)	*Factoring*

♦

Note: When an expression consists of several parts connected by addition and subtraction signs, each part (along with its sign) is known as a **term** of the expression. **Factors** are quantities that are multiplied together. For example, in the expression $3 + 5 \cdot 6$ the terms are 3 and $5 \cdot 6$. In the term $5 \cdot 6$, 5 and 6 are factors.

Example 2-2.4:   Factor $4x^2 - 8x$

Solution: We will look for a common factor in each term of the expression. The expression can be written as $4x(x) - 4x(2)$. There are two terms and a common factor is $4x$, so

$$4x^2 - 8x = 4x(x - 2).$$

♦

Activity 2.3

Factor the following expressions.

a) $7p - 35$ b) $2\pi h - 4\pi$

When multiplying two expressions we sometimes collect similar terms into one term. So when factoring we may have to "undo" this and split a term into several terms.

Example 2-2.5:   Multiply (x + 3)(x+2).

Solution: $(x+3)(x+2) = x(x+3) + 2(x+3) = x^2 + 3x + 2x + 6 = x^2 + 5x + 6$

$\blacklozenge$

Example 2-2.6:   Factor $a^2 + 3ab + 2b^2$.

Solution:
$$a^2 + 3ab + 2b^2 = a^2 + ab + 2ab + 2b^2 = a(a+b) + 2b(a+b) = (a+2b)(a+b).$$

$\blacklozenge$

Activity 2.4

Factor the following:

a) $a^2 + 5ab + 6b^2$

b) $1 + a + a^2 + a^3$

c) $1 + a + a^2 + a^3 + a^4 + a^5$

d) $1 + a + a^2 + a^3 + a^4 + a^5 + a^6 + a^7$

e)   Looking for a pattern in parts b - d, generalize the result. This means, factor $1 + a + a^2 + a^3 + a^4 + a^5 + a^6 + a^7 + \ldots + a^n$, where n is an odd integer.

Homework Section 2 - 2

1.   Mentally compute 297 + 1582 + 103, stating the properties you used.

2.   Mentally compute $150 \cdot 23 \cdot 6$, stating the properties you used.

3.   Mentally compute 1 + 2 + 3 + 4 + 5 + 6 + 7 + 8 + 9 + 10, stating the properties you used.

4.   Compute the sum of the first 99 counting numbers, 1+2+3+...+99.

5.   In the expression $x^2 - 2xy - 3y^2$, is x^2 a term or a factor?

6.   Name the factors in the expression $3x(x+y)$.

7.   Use the distributive property to determine how many additive terms would be in $(a+b+c+d+e)(x+y+z)$?

8.   Factor $1+a^2-a-a^3$.

9.   Factor $x^2 - 2xy - 3y^2$.

10.   Factor $(P-xP)+y(P-xP)$

11.   A math exam contained only two questions. Problem one was solved by 70% of the students. Problem two was solved by 60% of them. Every student solved at least one of the problems. Nine students solved both problems. How many students took the exam?

        Hint: Let n represent the number of students that took the exam. How many students answered problem 1 correctly? How many students only answered problem 1 correctly? How many students only answered problem 2 correctly? Add the number of students who only answered problem 1 to the number of students who only answered problem 2. What is the relationship to this sum and the total number of students who took the test?

Section 2-3: Order of Operations

Suppose you had to compute $2 + 5 \cdot 6$. You might perform the operations from left to right and get 42, whereas a friend of yours might first multiply the 5 and 6 and then add 2, to get an answer of 32. In mathematics, we cannot have any inconsistencies such as this. It is important that any expression involving numbers and operations represent exactly one number. There is a set of rules (**order of operations**) to prevent different answers from occurring. The order of operations has many names. For example, computer programmers usually use one of these terms: *order of precedence, math hierarchy,* or *order of computation.*

An understanding of the order of operations is fundamental in working with algebraic expressions like $5 + x^2$ and $(5 + x)^2$. These two expressions look similar, but the order of operations tells us they have very different meanings. Since most computer programs perform arithmetic calculations an understanding of the order of operations is essential to successful programming.

ORDER OF OPERATIONS

If **no** grouping symbols are present do the following:

1. Evaluate all powers, working from left to right.
2. Do any multiplications and divisions in the **order** in which they occur, reading from left to right.
3. Do any additions and subtractions in the **order** in which they occur, reading from left to right.

If grouping symbols are present, do the following:

1. First use the above steps within each pair of parentheses.
2. If the expression includes nested grouping symbols, evaluate the expression in the innermost set of parentheses first.

Grouping symbols may be ( ), [ ] , or { }.

Note: *Some people use the acronym "Please Excuse My Dear Aunt Sally" to remember the order of operations. Here the P stands for parentheses, E for exponent, M for multiplication, D for division, A for addition, and S for subtraction. This acronym can be very misleading for people because it can be misinterpreted to mean that multiplication must be performed before division and addition before subtraction. Remember, you do multiplication and division the order in which they appear, if a division appears before a multiplication you do division first. Similarly, if a subtraction appears before an addition, you do subtraction first.*

Example 2-3.1: Evaluate $7 + 3 \cdot 2 - 18 \div 6$ without using a calculator.

Solution: There are no grouping symbols or exponents, so do multiplications and divisions from left to right in the **order they appear**, then do additions and subtractions.

$$
\begin{aligned}
7 + 3 \cdot 2 - 18 \div 6 &= 7 + 6 - 18 \div 6 && \text{multiply 3 and 2} \\
&= 7 + 6 - 3 && \text{divide 18 and 6} \\
&= 13 - 3 && \text{add 7 and 6} \\
&= \mathbf{10} && \text{subtract 3 from 13}
\end{aligned}
$$

◆

Note: Often it is a good idea in complex arithmetic expressions to use grouping symbols even when they are not necessary. This can make the expressions easier to read. For example, the expression $7 + 3 \cdot 2 - 18 \div 6$ can be written in a more readable form using grouping symbols:

$7 + (3 \cdot 2) - (18 \div 6)$.

Example 2-3.2: Evaluate $6 + 3(8 - 3)$ without using a calculator.

Solution: According to the order of operations, we have to calculate what is inside the grouping symbols first:

$$
\begin{aligned}
6 + 3(8 - 3) &= 6 + 3(5) && \text{subtract 3 from 8} \\
&= 6 + 15 && \text{multiply 3 and 5} \\
&= \mathbf{21} && \text{add 6 and 15}
\end{aligned}
$$

◆

Example 2-3.3: Evaluate $6 - 4[8 + 3(7 - 3)]$ without using a calculator.

Solution: This expression has nested grouping symbols where the [ ] are used in the same way as parentheses. In a case like this, we move to the innermost set of grouping symbols first and begin simplifying:

$$
\begin{aligned}
6 - 4[8 + 3(7 - 3)] &= 6 - 4[8 + 3(4)] && \text{subtract 3 from 7} \\
&= 6 - 4[8 + 12] && \text{multiply 3 and 4} \\
&= 6 - 4[20] && \text{add 8 and 12} \\
&= 6 - 80 && \text{multiply 4 and 20} \\
&= \mathbf{-74} && \text{subtract 80 from 6}
\end{aligned}
$$

◆

Example 2-3.4: Evaluate $18 \div 3 \cdot 2 + 7 - 2 \cdot 4$ without using a calculator.

Solution: There are no grouping symbols or exponents, so we go from left to right and do all multiplications and divisions in the order they appear. In this case a division comes before a multiplication, so the division is done first.

$$18 \div 3 \cdot 2 + 7 - 2 \cdot 4 = 6 \cdot 2 + 7 - 2 \cdot 4 \qquad \text{divide 18 by 3}$$
$$= 12 + 7 - 8 \qquad \text{multiply 6 and 2,}$$
$$\text{multiply 2 and 4}$$
$$= 19 - 8 \qquad \text{add 12 and 7}$$
$$= \mathbf{11} \qquad \text{subtract 8 from 19}$$

$\blacklozenge$

Example 2-3.5:   Evaluate $9 \cdot 2^3 + 36 \div 3^2 - 8$ without using a calculator.

Solution: We do **exponents** (powers) first.

$$9 \cdot 2^3 + 36 \div 3^2 - 8 = 9 \cdot 8 + 36 \div 9 - 8 \qquad 2^3 = 2 \cdot 2 \cdot 2 = 8 \text{ and } 3^2 = 3 \cdot 3 = 9$$
$$= 72 + 4 - 8 \qquad \text{multiply 9 and 8, then}$$
$$\text{divide 36 by 9}$$
$$= 76 - 8 \qquad \text{add 72 and 4}$$
$$= \mathbf{68} \qquad \text{subtract 8 from 76}$$

$\blacklozenge$

When computing an expression such as $\dfrac{2 \cdot 5 + 8}{6 - 3}$, the fraction bar has the same function as grouping symbols. **Any operations that appear above or below a fraction bar should be completed first.**

Example 2-3.6:   Compute $\dfrac{2 \cdot 5 + 8}{6 - 3}$.

Solution: We have to compute $2 \cdot 5 + 8$ and $6 - 3$ before we divide.

$$\frac{2 \cdot 5 + 8}{6 - 3} = \frac{18}{3} = \mathbf{6}$$

In order to compute this example on a calculator or a computer, you would have to write this problem in a horizontal line.

So $\dfrac{2 \cdot 5 + 8}{6 - 3}$ becomes $(2 \cdot 5 + 8) \div (6 - 3)$ or $(2 \cdot 5 + 8) / (6 - 3)$.

$\blacklozenge$

Activity 2.5

Evaluate the following **without** using a calculator:

a) $12 \div 6 \cdot 2 - 7(3 - {}^{-}2)$

b) $4[6 - 4(2 - 7)]$

c) $5 \cdot 3^4 + 16 \div 8 - 2^2$

d) $\dfrac{8 \cdot 2 - 4}{10 - 4 \cdot 3}$

Homework Section 2 - 3

1. According to the order of operations, what is the last operation performed in the following expression?

$$3(4+6\cdot 4)-12+16$$

2. Evaluate the following **without** using a calculator:

   a) $12\div 4\cdot 3-2(3+9\cdot 4)$

   b) $8\cdot 2-10\div 5\cdot 2+3$

   c) $4^2-(5-2\cdot 3)^3$

   d) $\dfrac{4\cdot 2+52}{3\cdot 2}$

3. Insert parentheses so that when the following expression is evaluated the answer is 16.

$$6\cdot 3-70+4-60\div 2+5$$

<u>***Section 2-4: Using Excel to Calculate***</u>

When using a calculator or Excel to evaluate an expression such as $\dfrac{1.2+5.7}{0.3\cdot 2.3}$, many people separately evaluate the **numerator** (top) and the **denominator** (bottom). They write these two answers down getting a final answer by dividing the two. If you know how to properly use grouping symbols and apply the order of operations, the answer can be obtained by entering a continuous succession of keystrokes. In some cases doing calculations separately can lead to rounding errors. As an example consider the following problem:

> If you travel from A to B at an average speed of 65 mph and return from B to A along the same route at an average speed of 55 mph, what is your average speed for the round trip?

> To solve this problem you can use the following formula:

$$\textbf{average speed} = \dfrac{2}{\dfrac{1}{r}+\dfrac{1}{s}}\text{, where r is the average speed from A to B}$$

> and s is the average speed from B to A.

One of our goals in this unit, is to be able to perform the calculation

$$\dfrac{2}{\dfrac{1}{65}+\dfrac{1}{55}}$$

on a calculator or Excel in a single series of keystrokes. If you do the calculation by first dividing 1 by 65, rounding that answer to three decimal places, then dividing 1 by 55, rounding to three decimal places, adding the two and then dividing 2 by that answer, you get an average speed of 60.61 mph. Doing the calculation in a single series of keystrokes gives an average speed of 59.58 mph, which is a significant difference. However, if you write the individual calculations with many decimal places you significantly reduce the rounding error.

In order to use a spreadsheet or program a computer in a language such as Java, it is essential to be able to write the above formula on a horizontal line which is the same way you would enter it in Excel or on your calculator. It would look like 2/(1/65 + 1/55). Knowing the order of operations is essential to being able to write this expression correctly.

Example 2-4.1: Write the following expressions horizontally.

a) $\left(\dfrac{343}{343-27}\right)(400)$

b) $\dfrac{255-(0.500)(40.0)(9.80)}{40.0}$

c) $\dfrac{a+b-c}{2ac}$

Solution: a) $(343 / (343 - 27))(400)$

The grouping symbols around 400 are not required but may make the expression easier to read.

b) $(255 - (0.500)(40.0)(9.80)) / 40.0$

c) $(a + b - c) / (2ac)$

In this example, make sure you understand why the grouping symbols are necessary around *2ac*. ◆

Activity 2.6

Write the following expressions horizontally.

a) $$\dfrac{49 + 14}{49 + 7 + 14}$$

b) $$\dfrac{\dfrac{3}{5}}{\dfrac{4}{7}}$$

c) $$\dfrac{f}{1 - r} - s - f$$

Now that you have had practice writing expressions horizontally, let's calculate some expressions using Excel. When using Excel or a calculator, you have to remember that the minus sign is used in two different ways:

(1) to indicate subtraction and

(2) to designate a negative number.

Example 2-4.2: Compute 5(17 + 53) using Excel.

Solution: You want to enter the expression the way it is written, except that Excel does not use implicit multiplication. You have to use * to indicate that (17 + 53) is multiplied by 5. **Whenever you are going to do a calculation start with an equal sign (=)**, and then enter the expression. As you can see on the top of the next page, the expression has been entered into cell A1. Once you enter =5*(17+53) hit the enter key and the answer will appear in cell A1.

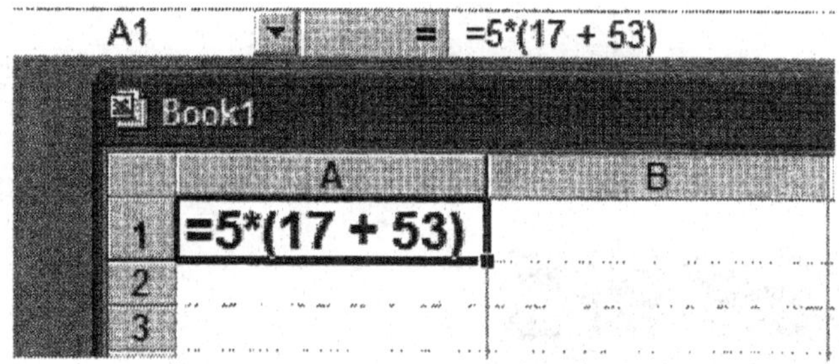

Example 2-4.3: Compute $\dfrac{4+8}{2\cdot 3}$ using Excel.

Solution: We have to rewrite the expression horizontally. Remember the fraction bar indicates grouping the numerator and the denominator. The expression really looks like $\dfrac{(4+8)}{(2\cdot 3)}$. Writing this horizontally we have $(4+8)/(2\cdot 3)$. Entering this in Excel we have the following:

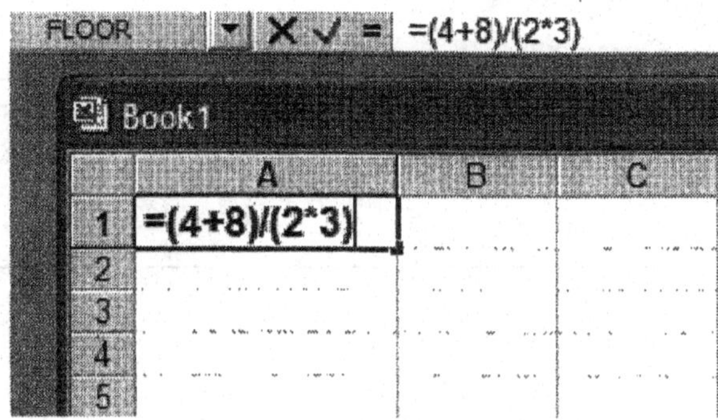

When we hit the enter key we get

Activity 2.7

What number will be printed when Enter is hit in the following worksheets?

a)

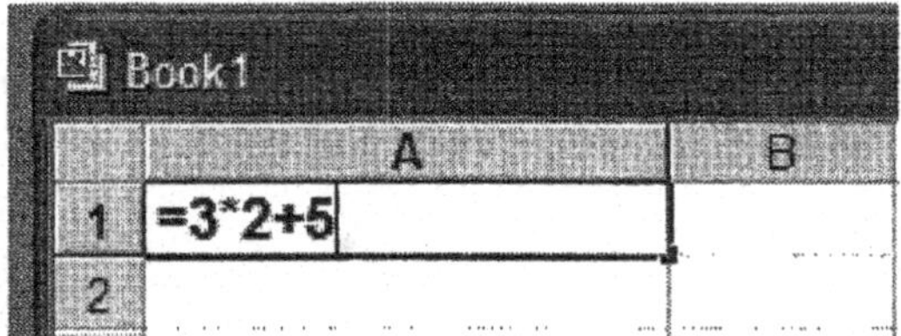

b)

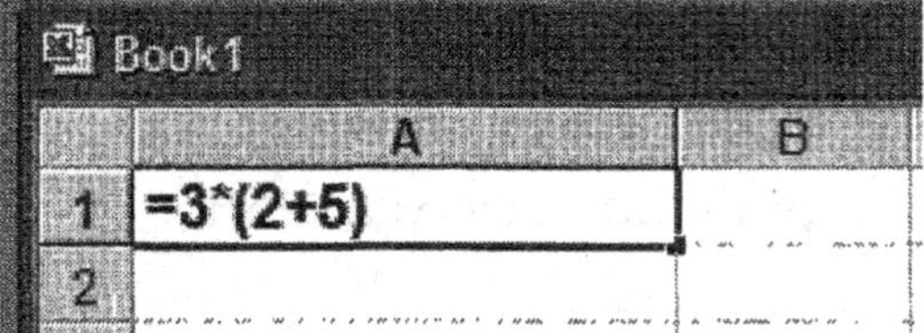

c)

Activity 2.8

Calculate the following problems using Excel. Where necessary, round answers to three decimal places.

    a) $1.2(4.5 - 8.4)$ b) $-3.2(4.6 - {}^- 5.9)$

    c) $\dfrac{123(40.2)}{24.65 - 12.93}$

Homework Section 2 - 4

1. Write the following expressions horizontally.

    a) $$\frac{5}{3 \cdot 2} - 8$$

    b) $$\frac{4 - 6 \cdot 4}{7}$$

    c) $$\frac{q^n - 1}{q - 1}$$

2. What number will be printed when Enter is hit in the following worksheets?

    a)

	A
1	=3^2-15/3
2	

    b)

	A
1	=4/2-6*(4/2+3)

    c)

	A
1	=3+9*2^2+1

3. Most people work with decimals, but sometimes you may want to do some arithmetic with fractions. In this exercise you will learn how to work with numbers like $4\tfrac{7}{8}$ using Excel.

    a) Enter the text shown below into Excel.

	A	B	C	D	E	F	G	H
1	**Computing with Fractions**							
2								
3	**Adding fractions to give an answer to the nearest 1/2**							

b)   Enter the numbers in A4:C4. Typing the 3 followed by a space and then 2/3 enters the number in A4. Look in the formula bar and 3.666667 is displayed. The ¾ is entered as 0 ¾; if you only enter ¾, Excel will give you a date.

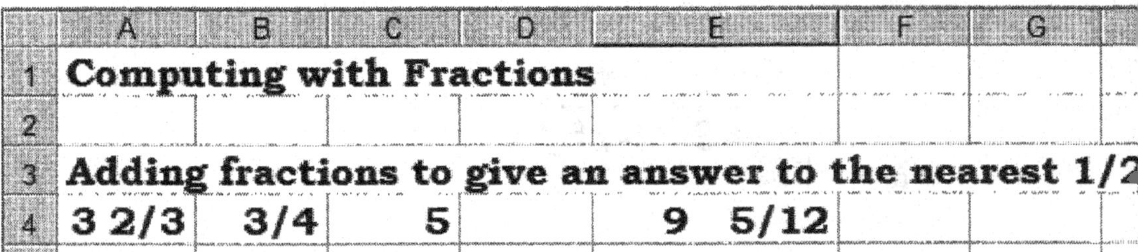

c)   In cell E4 enter **=A4+B4+C4**. The result is displayed as 9 5/12.

We wanted the answer rounded to the nearest ½, so we need to format cell E4. Use *Home/Number* launcher; (southeast pointing triangle next to the word Number). This opens up the formatting dialog from where you can select *As Halves (1/2)* in the Fraction category.

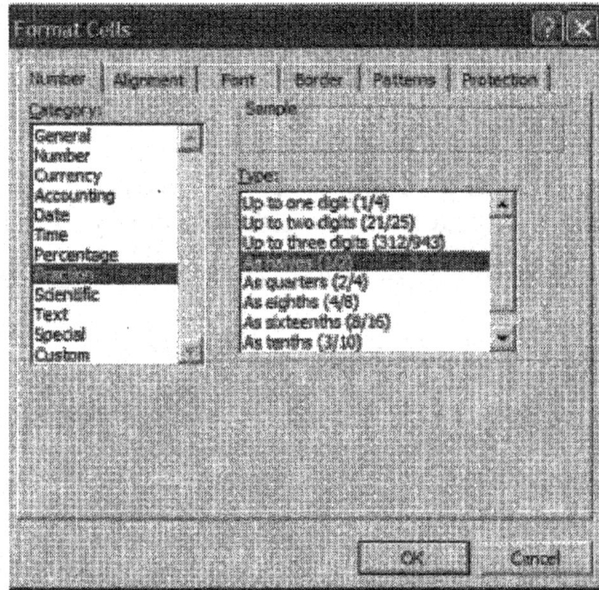

The result is now displayed as 9 1/2.

	A	B	C	D	E	F	G	
1	**Computing with Fractions**							
2								
3	**Adding fractions to give an answer to the nearest 1/2**							
4	3 2/3	3/4	5		9 1/2			

d)   Use Excel to compute the following.

$$\frac{3}{7} - \frac{6}{17}$$

Section 2-5: Symbolic Expressions

In Section 2 we used symbols to state the commutative, associative, and distributive properties of numbers. We can also use formulas for addition, subtraction, multiplication, and division of fractions. Let $\dfrac{k}{a}, \dfrac{m}{b}$ be arbitrary rational numbers, where k, a, m, and b are integers, and $a \neq 0, b \neq 0$, then

$$1. \quad \frac{k}{a} \pm \frac{m}{b} = \frac{kb \pm ma}{ab}$$

$$2. \quad \frac{k}{a} \cdot \frac{m}{b} = \frac{km}{ab}$$

$$3. \quad \frac{k}{a} \div \frac{m}{b} = \frac{kb}{ma}$$

$$4. \quad \frac{k}{a} = \frac{kb}{ab} \quad \textit{(equivalent fractions)}$$

These formulas can be used for any specific values of $\dfrac{k}{a}, \dfrac{m}{b}$.

Example 2-5.1:   Add $\dfrac{2}{3} + \dfrac{4}{7}$

Solution: In this example $k = 2$, $a = 3$, $m = 4$, and $b = 7$. Using the first formula above, we get

$$\frac{2}{3} + \frac{4}{7} = \frac{2 \cdot 7 + 4 \cdot 3}{3 \cdot 7} = \frac{14 + 12}{21} = \frac{26}{21} \qquad \blacklozenge$$

Let's use Excel to add $\dfrac{2}{3}$ and $\dfrac{4}{7}$. We will just use the formula for adding fractions, $\dfrac{k}{a} + \dfrac{m}{b} = \dfrac{kb + ma}{ab}$, and substitute values for k, m, a, and b. In this example, $k = 2$, $m = 4$, $a = 3$, and $b = 7$. An example of an Excel worksheet, which will add two fractions, is given on the next page. In cell C9 the formula appears, notice the grouping symbols that have to be used when writing $\dfrac{kb + ma}{ab}$ horizontally. Also remember that when you enter the formula in C9 and hit enter you will see **1.23809524** instead of the formula.

	A	B	C
1			
2	**Adding Fractions**		
3			
4		k =	2
5		m =	4
6		a =	3
7		b =	7
8			
9		k/a + m/b =	=(C4*C7+C5*C6)/(C6*C7)

Activity 2.9

Modify the spreadsheet above to compute the following.

a) $\dfrac{2}{3} \cdot \dfrac{4}{7}$

b) $\dfrac{2}{3} \div \dfrac{4}{7}$

Activity 2.10

Use $\dfrac{k}{a} \pm \dfrac{m}{b} = \dfrac{kb \pm ma}{ab}$ to do the following problems.

a) $\dfrac{2}{x} + \dfrac{3}{y}$

b) $\dfrac{1}{x} - \dfrac{2}{x-3}$

Example 2-5.2: $\dfrac{k}{a} = \dfrac{kb}{ab}$ allows you to find equivalent fractions. For example, to find two fractions equivalent to $\dfrac{4}{7}$, we can do the following:

$$\frac{4}{7} = \frac{4 \cdot 3}{7 \cdot 3} = \frac{12}{21}$$

$$\frac{4}{7} = \frac{4 \cdot 2}{7 \cdot 2} = \frac{8}{14}$$

Here we multiplied the numerator and denominator of $\dfrac{4}{7}$ by 3 and by 2 to get two equivalent fractions.

Example 2-5.3: Evaluate $3x(x - \dfrac{3}{7})$ when $x = \dfrac{2}{9}$.

Solution: In order to enter this in Excel, we have to be able to write the expression horizontally. In addition, we need to make the proper use of grouping symbols. We would enter the following in cell A1 and then hit **Enter**:

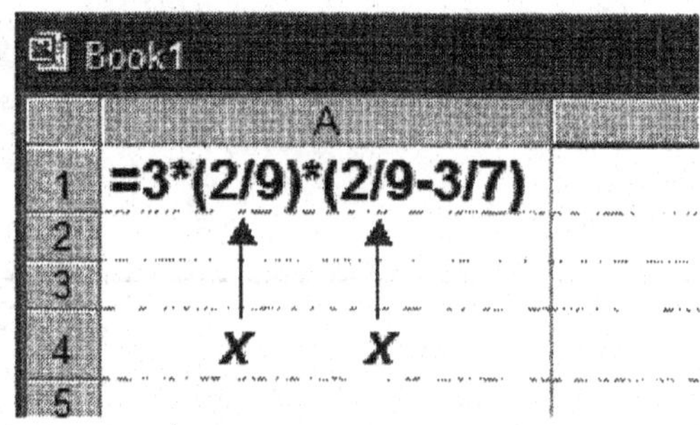

The answer **-0.137566138** will appear.

Remember, using grouping symbols for complex expressions, even when they are not needed, can make the expression easier to read. The above could have been entered in Excel as follows:

$$=3*(2/9)*((2/9)-(3/7))$$

$\blacklozenge$

You could also have set up the following Excel worksheets where the value of x is entered in B1, **=3*B1*(B1-3/7)** is entered in C3, and **=B1** is entered in E3. The advantage is that if the expression had to be evaluated for different values of x, you just have to enter the value in B1. The second worksheet uses decimals rather than fractions.

	A	B	C	D	E
1	x =	2/9			
2					
3	3x(x-3/7)	is equal to	- 26/189	when x =	2/9

	A	B	C	D	E
1	x =	0.22222222			
2					
3	3x(x-3/7)	is equal to	-0.137566138	when x =	0.22222222

Activity 2.11

Use Excel to evaluate x(4x - 30), when x = $\dfrac{2}{3}$.

Example 2-5.4: The **circumference** of a circle (distance around) is given by the formula C = 2πr, where r is the radius of the circle. Find the circumference of a circle if the radius is 6.251 inches. Round your answer to 3 decimal places.

Solution: Instead of 3.14 for π we will use Pi( ) which gives an approximation to π to 14 decimal places. Also notice that in the Excel worksheet below we made it general. Any time cell C3 is changed; cell C5 gives a new value for the circumference.

	A	B	C	D	E
1	**Circumference of a Circle**				
2					
3		radius =	6.251		← Input
4					
5		Circumference =	=2*PI()*C3		
6					

When **Enter** is hit, the circumference will be **39.276** inches.

♦

Activity 2.12

The area of a circle is given by the formula $A = \pi r^2$, where r is the radius of the circle. Use Excel to find the area of a circle whose radius is 2.5 cm. Round your answer to 3 decimal places.

Homework Section 2-5

1.   Using the formulas at the beginning of the section, compute the
     following:

    a) $\dfrac{3}{5}+\dfrac{4}{9}$ b) $\dfrac{4}{x}-\dfrac{1}{2y}$

    c) $\dfrac{4}{5}\cdot\dfrac{2}{3}$ d) $\dfrac{4}{5}\div\dfrac{2}{3}$

2.   Write three fractions equivalent to $\dfrac{3}{11}$.

3.   Use Excel to evaluate $\dfrac{a-b}{b+c}$ when

    a)   a = 5, b = 7, and c = 9

    b)   a = 1/3, b = 4/5, and c = 1/2

<u>**Section 2-6: Formulas Involving Subscripts**</u>

In most science courses, instead of using different letters to represent different quantities, the same letter is used with different **subscripts**. Just as a and b in an expression or formula represent different quantities, so do R_1 and R_2. Look at the nutritional value of a particular food and you will see vitamins such as Vitamin B_6 and Vitamin B_{12} identified by subscripts.

Example 2-6.1:   Evaluate $2a_1 - 3a_2 + a_1 a_2$ when $a_1 = -4$ and $a_2 = 5$.

Solution: Substituting - 4 for a_1 and 5 for a_2 gives us 2(- 4) - 3(5) + (- 4)(5).

Using Excel, your worksheet could look like the following:

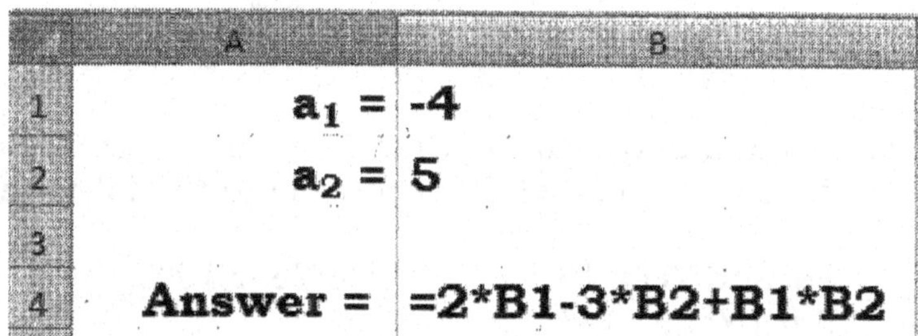

The answer in cell B4 will be **-43**.   ◆

Note: It is not necessary, but in the last example A1 was displayed as a_1 with a subscript. To do this use *Home/Font/Font Launcher* and check the subscript box in the Format Dialog.

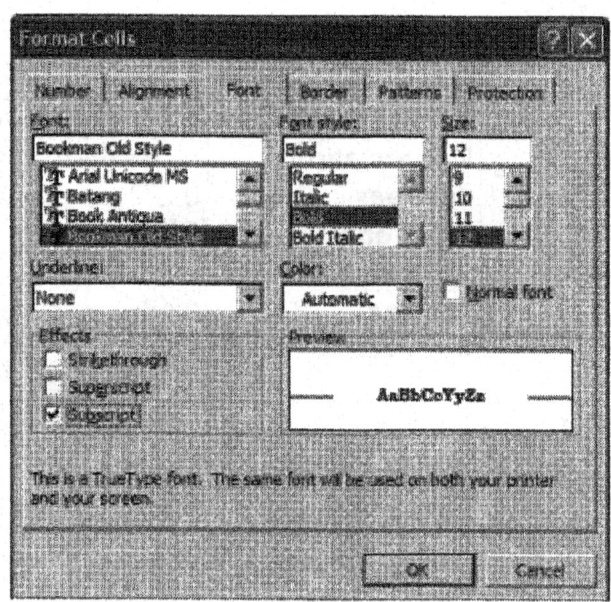

Activity 2.13

The equivalent resistance R (in ohms) of two resistors connected in parallel is given by the formula $R = \dfrac{R_1 R_2}{R_1 + R_2}$, where R_1 and R_2 are the resistances of the two resistors (in ohms). Use Excel to find the value of R if R_1 = 15,000 ω and R_2 = 3,000 ω.

Activity 2.14

According to Edmunds.com, a 2009 Honda Civic DX 2dr Coupe declines in value during the first years that you own it, according to the following table:

First Year   27%   Second Year 13%   Third Year 13%
Fourth Year   14%   Fifth Year   14%

1. Calculate by hand, and then set up an Excel worksheet similar to the one below to answer the following questions.

	A	B	C	D
1	**Car Depreciation**			
2				
3		**Original Amount**	15305	
4				Input values are shaded
5		**Depreciation**		
6		**1st Year**	0.27	
7		**2nd Year**	0.13	
8		**3rd Year**	0.13	
9		**4th Year**	0.14	
10		**5th Year**	0.14	
11				
12	**Worth of Car After**			
13	I Year	=C3-C6*C3		
14	2 Years	=B13-C7*B13		
15	3 Years	=B14-C8*B14		
16	4 Years	=B15-C9*B15		
17	5 Years	=B16-C10*B16		

a) How much would the car be worth after one year, if you bought it for $15,305?

b) How much is the car worth after 5 years?

2.   You can generalize the process by letting
     V = present value, P = purchase price, d_n = depreciation percent in
     nth year

     The generalization then becomes

     $$V = P(1 - d_1)(1 - d_2)...(1 - d_n)$$

     Use this formula to find the value of the Honda Civic after three years.

3.   To calculate the worth of the Honda after one year, you take 27% of
     $15,305, which is $4132.35, and then subtract $4132.35 from $15,305.
     How could you calculate the worth after one year in just one step?

Unit 2 Review Exercises

1. Evaluate the following expressions with paper and pencil.

   a) $3 \cdot 40 \div 12 \cdot 72 \div 24$ b) $3(45 - 21) + 7 - 10$

   c) $(45 - 21)(7 - 10)$ d) $64 - 5[14 + 20(14 - 3)]$

   e) $\dfrac{3(12 - 7) - 3}{4 + 8}$ f) $\dfrac{24 - 4(3 \cdot 5 - 2)}{2 \cdot 5} - 7$

   g) $\dfrac{11 - 5(3)}{7(3) - 10(-3)}$ h) $4\left(6 - \dfrac{3 - 6(2)}{3}\right)$

   i) $123 - 2\{18 + 2(19 - 2[7 + 1])\} - 4$

   j) $27 \div 3^2 + 12 - 9 + 3$

2. Write parts e, f, g, and h above horizontally.

3. Use Excel to evaluate the following expressions, using $a_1 = 2$, $a_2 = 6$, and $a_3 = 8$.

   a) $1.5a_1 + a_2 - 6a_3$ b) $a_3 - a_1 \cdot a_2$

   c) $\dfrac{5a_2 + 6a_1}{a_1 \cdot a_3 - 2}$ d) $5\{a_2 - a_1[4 - (3 + a_3)] + 3\}$

   e) $a_1 a_3 - a_2(a_1 - a_3)$

4. Given the expression $5x^3 + 2$, which of the following are equivalent to this expression?   (There may be more than one correct answer.)

   a)   5 * x * x * x + 2   b)   5 * x * x * (x + 2)

   c)   (5 * x) * x * (x + 2)   d)   5 * (x * x * x) + 2

   e)   5 * x * (x * x * 7)

5.   Factor the following expressions.

a) $18 - 6x^2$ b) $6d^2 - 14d$

c) $x^2 + xy - 2y^2$

6.   Name the terms in each of the following expressions.

a) $2\pi r$ b) $-\frac{1}{2} gt^2 + v_0 t + h_0$

7.   Using the following depreciation schedule of 30%, 20%, 15%, 12%, and 10%, calculate the value after 5 years of use of a car that had an original price of $30,000.

8.   Using the depreciation schedule in the previous problem, suppose a three-year old car is presently valued at $15,000, what was its original price?

9.   Show that $\dfrac{k}{a} + \dfrac{m}{b} = \dfrac{kb + ma}{ab}$.

10.   Compute the following:

a) $\dfrac{4}{11} + \dfrac{3}{4}$ b) $\dfrac{4}{11} \div \dfrac{3}{4}$

c) $\dfrac{1}{x} - \dfrac{x+1}{x-2}$ d) $\dfrac{4}{x^2} + \dfrac{x-1}{3}$

11.   Use the fact that $\dfrac{k}{a} = \dfrac{kb}{ab}$ to simplify the following fractions.

a) $\dfrac{14}{20}$ b) $\dfrac{a}{a^2}$ c) $\dfrac{2b}{b^2 + 5b}$

<u>**Section 3-1: Introduction**</u>

Recall that multiplication can be thought of as a shorthand notation for repeated addition, $3 \cdot 4 = 4 + 4 + 4$. In a similar way exponents are a shorthand notation for repeated multiplication, $2^3 = 2 \cdot 2 \cdot 2$. As with any special notation you have to learn to read and use it correctly. Let's look at the definition that follows.

<table>
<tr>
<td>

Exponential Notation

</td>
<td>

Let x be a real number and n be a positive integer. x used as a factor n times

$$(x)(x)(x) \cdots (x)$$
$$\text{n times}$$

is called

       x to the nth power

or

       x to the nth

or

       x raised to the n.

We call n the ***exponent*** or power and x the ***base***, and write $(x)(x)(x) \cdots (x)$ simply as $\mathbf{x^n}$.
$\qquad\qquad\qquad \text{n times}$

For example, 5^3 would be read "5 to the 3rd power", "5 to the 3rd", or "5 raised to the 3rd". The **base** is 5 and the **exponent** is 3.

</td>
</tr>
</table>

Section 3-2: Base of an Exponent

When dealing with exponential notation it is very important to determine what the base of an exponent is. **As a general rule, the exponent applies to the symbol that precedes it**. For example, the symbol that precedes the exponent in 5^3 is 5, so the base for the exponent is 5. This means that 5 is used as a factor 3 times, $5^3 = (5)(5)(5)$. Now consider $(4x)^2$, here the symbol that precedes the exponent is parentheses. This tells us that everything inside the parentheses is the base, so here 4x will be used as a factor twice, $(4x)^2 = (4x)(4x)$. **Remember that when parentheses are used, it is the entire quantity inside the parentheses that is repeated as the factor.**

Activity 3.1

Name the base for each exponent.

a) $(-5)^2$ b) -5^2

c) $(x + y)^2$ d) $4x^3$

Make sure that you understand the difference between parts a and b above. In part a, the base for the exponent is -5 because parentheses precede the exponent and the base is everything inside the parentheses. Therefore, $(-5)^2$ means $(-5)(-5)$. In part b, the base for the exponent is just 5, the symbol that precedes it. Since the base is just 5, 5 is all that gets squared. Thus -5^2 means $-(5 \cdot 5)$.

Note: It may be helpful to think of -5 as $-1 \cdot 5$. So in the problem above, -5^2 means $-1 \cdot 5^2$.

Activity 3.2

a)   Name the base for the exponent in the expression

$$4 + 5\{3 - 3(12 \div 6 \cdot 3 - 5 + 6)\}^3$$

Note: In the above expression, { }, are just grouping symbols. They are sometimes used to help read an expression. You do not enter them on your calculator, use ( ) instead.

b) $-x^3$ is the same as which of the following:

   i) $-x \cdot -x \cdot -x$ ii) $-1 \cdot x^3$ iii) $-(x \cdot x \cdot x)$

You have to be careful evaluating exponents using Excel. The key is to determine the base for the exponent. For example, to evaluate -5^2 you have to specify the base for Excel. If you enter =-5^2, Excel will output 25 because it assumes the base is -5. You will have to enclose the base with parenthesis.

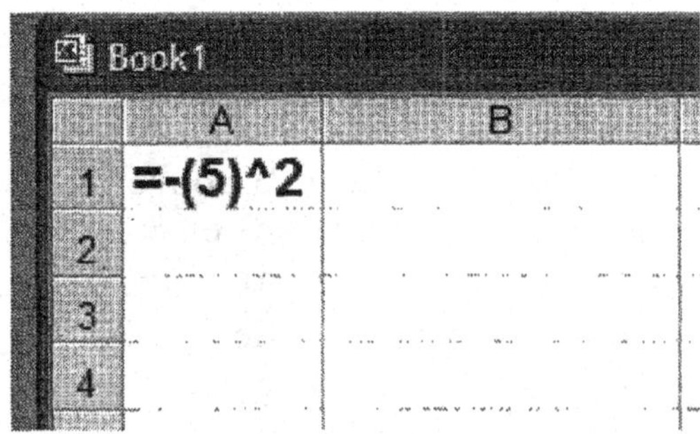

You could also have evaluated -5^2 using the Power function. In Excel it would be entered as follows:

<u>**Example 3-2.1**</u>:   Use Excel to evaluate the following expression. Show the sequence of keys you would use.

$$\frac{5+3^4}{39+4}$$

Solution: $\dfrac{5+3^4}{39+4}=2$

Notice you have to put grouping symbols around the numerator and the denominator.

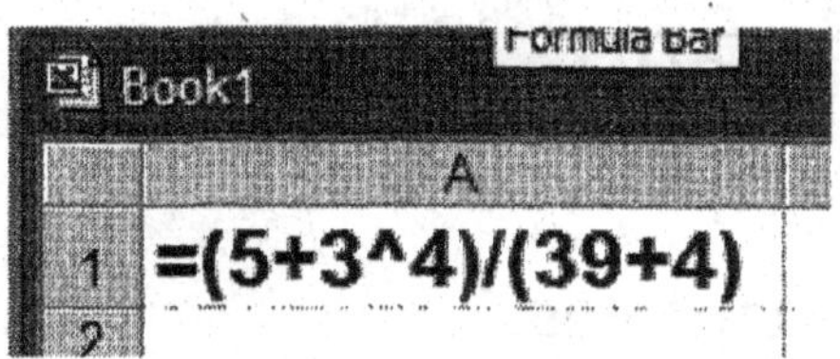

♦

Activity 3.3

Evaluate the following using Excel. Show the sequence of keys that you would use.

a) $3 + 5^2$

b) $(3 + 5)^2$

c) $^-6^2 - (3 - 4 \cdot 6)$

d) $\left(\dfrac{2-5^2}{2 \bullet 5}\right)^2$

e) $5 \cdot (-18)^3$

Example 3-2.2: Use Excel to perform the following calculation. Evaluate $1000\left(1+\dfrac{r}{4}\right)^{8}$ when $r = 0.05$. Round the answer to three decimal places.

Solution: First substitute the given value for the variable, $1000\left(1+\dfrac{0.05}{4}\right)^{8}$. Remember, you have to enter this expression horizontally.

	A
1	=1000*(1+0.05/4)^8
2	

The output will be **1104.486.** ◆

A better way to evaluate the formula in the last example would be to make your worksheet more general. Your worksheet may look like the following:

	A	B
1		
2	r =	0.05
3	Answer =	=1000*(1+B2/4)^8
4		
5		
6		

Value of r is in cell B2

The advantage of this worksheet is that now to evaluate the formula for different values of r, you just have to change cell B2 instead of retyping the whole formula.

Activity 3.4

Use Excel to perform the indicated calculations. Round off answers to three decimal places.

a)   Evaluate $(1 + t^3)^{27}$ when $t = 0.3$.

b)   Evaluate $-x^3 - 4x + 12$ when $t = -2.3$.

Homework Section 3-2

1.   Compute:

    a) 10^3 b) 10^7

2.   How many digits do you need to write 10^{1000} ?

3.   Name the base for each exponent.

    a) -7^6 b) $x+y^3$ c) $\left(\dfrac{5}{x+y}\right)^4$

4.   Evaluate the following using Excel. Round your answer to 3 decimal places.

    a) $\dfrac{2+7^3}{(6-9)^5}$ b) $4.3\left(\dfrac{2+8}{1-5^3}\right)^2$

5.   Use Excel to evaluate $1000\left(1+\dfrac{r}{12}\right)^{(5\cdot12)}$ when

    a) $r=0.05$ b) $r=0.035$ c) $r=0.0245$

    Round your answer to two decimal places.

Section 3-3: Zero and Negative Exponents

Let's look at the following sequence of numbers

$$256,\ 128,\ 64,\ 32,\ 16,\ 8,\ 4,\ 2.$$

In this sequence each number is half the number that precedes it. If we continue this sequence we get:

$$256,\ 128,\ 64,\ 32,\ 16,\ 8,\ 4,\ 2,\ 1,\ \frac{1}{2},\ \frac{1}{4},\ \frac{1}{8},\ \frac{1}{16},\ \dots$$

The sequence 256, 128, 64, 32, 16, 8, 4, 2 could be written as 2^8, 2^7, 2^6, 2^5, 2^4, 2^3, 2^2, 2^1. The pattern suggests the following continuation:

$$256,\ 128,\ 64,\ 32,\ 16,\ 8,\ 4,\ 2,\ 1,\ \frac{1}{2},\ \frac{1}{4},\ \frac{1}{8},\ \frac{1}{16},\ \dots$$

$$2^8,\ 2^7,\ 2^6,\ 2^5,\ 2^4,\ 2^3, 2^2, 2^1,\ 2^0,\ 2^{-1},\ 2^{-2},\ 2^{-3},\ 2^{-4},\ \dots$$

When we talked about exponents before we said that 2^4 meant *"2 used as a factor four times"*. You can see that this logic breaks down for 2^{-1}. What does it mean for *"2 to be used as a factor −1 times"*? It is an agreement that 2^{-n} (for positive integer n) means $\dfrac{1}{2^n}$.

Activity 3.5

How should the pattern continue for the given sequence?

$$27,\ 9,\ 3,\ 1,\ \frac{1}{3},\ \frac{1}{9},\ \frac{1}{27},\ \dots$$

$$3^3,\ 3^2,\ 3^1,\ \underline{\ \ \ },\ \underline{\ \ \ },\ \underline{\ \ \ },\ \underline{\ \ \ }$$

This leads us to a general definition of exponents.

$$\boxed{\begin{aligned}
&\textbf{Definition: } \text{For positive integers n and } b \neq 0, \\[2mm]
&\quad b^n = b \cdot b \cdots b \quad \text{(b used as a factor n times)} \\[2mm]
&\quad b^0 = 1 \\[2mm]
&\quad b^{-n} = \frac{1}{b^n}
\end{aligned}}$$

This notation allows us to write an expression such as $3 \cdot x \cdot x \cdot x \cdot y \cdot y \cdot z \cdot z \cdot z \cdot z \cdot z \cdot z$ in the more compact form $3x^3 y^2 z^6$.

Example 3-3.1: Rewrite the expression $\dfrac{2 \cdot a \cdot a \cdot a \cdot a \cdot a \cdot c}{b \cdot b \cdot b \cdot b \cdot b \cdot d \cdot d}$ in a more compact form.

Solution:
$$\frac{2 \cdot a \cdot a \cdot a \cdot a \cdot a \cdot c}{b \cdot b \cdot b \cdot b \cdot b \cdot d \cdot d} = \frac{2a^6 c}{b^5 d^2} = 2a^6 b^{-5} cd^{-2}$$

◆

Activity 3.6

a. Write a compact form for the following expressions:

    i) $a \cdot a \cdot a \cdot a \cdot a \cdot a \cdot a \cdot a \cdot a \cdot b \cdot b \cdot b$

    ii) $\dfrac{3 \cdot x \cdot x \cdot x \cdot x \cdot x}{b}$

b. Rewrite the following expressions using only positive exponents:

    i) $3ab^{-4}$

    ii) $a^{-4} b^{-1}$

    iii) $4a^{-4} b^2 c^{-2}$

Homework Section 3-3

1.   Write the following as decimals:

   a) 10^{-2} b) 10^{-4}

2.   Rewrite the following expressions using only positive exponents:

   a) $-3x^{-2}y^{5}$ b) $(2xy)^{-1}$

3.   Use Excel to evaluate $3 \cdot 4^{t}$ when t equals:

   a)   0   b)   -1   c)   -2   d)   -3   e)   -4

<u>**Section 3-4: Rules of Exponents**</u>

How do we multiply b^m by b^n if m and n are positive?

Example 3-4.1:   Multiply b^4 by b^3.

Solution: $b^4 \cdot b^3 = \underbrace{(b \cdot b \cdot b \cdot b)}_{4\ times} \cdot \underbrace{(b \cdot b \cdot b)}_{3\ times} = b^7$ ◆

This example suggests that in general, $b^m \cdot b^n = b^{m+n}$ (b^m is b used as a factor m times and b^n is b used as a factor n times). It turns out that the rule is valid if the exponents are negative, too.

Example 3-4.2:   Multiply b^4 by b^{-2}.

Solution: Using the rule above, $b^4 \cdot b^{-2} = b^{4+(-2)} = b^2$.

Let's check this answer. By definition,

$$b^4 \cdot b^{-2} = b^4 \cdot \frac{1}{b^2} = \frac{b \cdot b \cdot b \cdot b}{b \cdot b} = b^2 .$$ ◆

Activity 3.7

Multiply the following. Answers should contain positive exponents only.

a.) $7^5 \cdot 7^9$

b) $(-6)^9 \cdot (-6)^{-5}$

c) $x^{-7} \cdot x^4$

Let's see what happens when we calculate $(b^m)^n$. It may be helpful to try a couple of examples and then try and generalize our results.

Example 3-4.3:   Find $(b^2)^3$.

Solution: $(b^2)^3 = \underbrace{b^2 \cdot b^2 \cdot b^2}_{3\ times} = (b \cdot b) \cdot (b \cdot b) \cdot (b \cdot b) = b^6$

So, $(b^2)^3 = b^6$.   ◆

Let's try an example with different signs.

Example 3-4.4: Find $(2^{-3})^4$.

Solution:

$$(2^{-3})^4 = \underbrace{(2^{-3})\cdot(2^{-3})\cdot(2^{-3})\cdot(2^{-3})}_{4\,times} = \frac{1}{2\cdot2\cdot2}\cdot\frac{1}{2\cdot2\cdot2}\cdot\frac{1}{2\cdot2\cdot2}\cdot\frac{1}{2\cdot2\cdot2} = \frac{1}{2^{12}} = 2^{-12}$$

So, $(2^{-3})^4 = 2^{-12}$. ♦

These last two examples suggest that

$$(b^m)^n = b^{mn}.$$

Activity 3.8

Use the formula above to simplify the following. Answers should contain positive exponents only.

a) $\left(x^4\right)^5$

b) $\left(y^{-4}\right)^2$

Activity 3.9

Find a formula for $\dfrac{b^m}{b^n}$. Try looking at specific examples first.

Again going from the specific to the general, we can show the formula $(ab)^n = a^n b^n$.

Example 3-4.5: Study the following computation:

$$(2 \cdot 3)^4 = (2 \cdot 3)(2 \cdot 3)(2 \cdot 3)(2 \cdot 3)$$
$$= (2 \cdot 2 \cdot 2 \cdot 2)(3 \cdot 3 \cdot 3 \cdot 3) \quad \text{(commutative property)}$$
$$= 2^4 \cdot 3^4$$

Making this process more general,

$$(ab)^n = \underbrace{(ab)(ab)(ab) \cdots (ab)}_{n \ times}$$
$$= \underbrace{(a \cdot a \cdot a \cdots a)}_{n \ times}\underbrace{(b \cdot b \cdot b \cdots b)}_{n \ times}$$
$$= a^n b^n$$

◆

Activity 3.10

Use the formula above to simplify the following. Answers should contain positive exponents only.

a) $(2xy)^3$

b) $(3x)^{-2}$

Activity 3.11

Find a formula for $\left(\dfrac{a}{b}\right)^n$.

Here is a summary of the rules for exponents.

Rules of Exponents: For m and n any integers and $b \neq 0$, $a \neq 0$:

1. $b^m \cdot b^n = b^{m+n}$

2. $\dfrac{b^m}{b^n} = b^{m-n}$

3. $(b^m)^n = b^{mn}$

4. $(ab)^n = a^n b^n$

5. $\left(\dfrac{a}{b}\right)^n = \dfrac{a^n}{b^n}$

6. $b^{-n} = \dfrac{1}{b^n}$

7. $b^0 = 1$

Homework Section 3-4

1. Use the Rules of Exponents to simplify the following. Write all answers with positive exponents.

   a) $\left(x^2 y^{-1}\right)^2$ b) $x^3 \cdot x^{-5}$ c) $\dfrac{xy^7}{y^3}$

2. The point of this section has been to introduce rules for exponents, which allow us to write exponential expressions in different forms. These skills will be important when we study exponential functions in a later unit.

   a) Rewrite 4^{3t} so that the base is 2.

   b) Rewrite 4^{3t} so that the base is 8.

3. Give an explanation of why the rule $a^n \cdot a^m = a^{n+m}$ holds.

4. Show that $4(x+4)^2 = (2x+8)^2$ using your knowledge of factoring and the rules of exponents.

5. Does $(-a)^{973}$ equal a^{973} or $-a^{973}$? Support your answer.

<u>*Section 3-5: Scientific Notation*</u>

Many quantities in science are very large or very small and the ordinary way of writing them becomes awkward. For instance, in one gram of the element hydrogen there are **6,023,000,000,000,000,000,000,000** atoms of hydrogen. The diameter of the average red corpuscle cell is **0.00008** cm. **Scientific notation** is one way scientists deal with these extreme numbers. This method is now needed to describe the size of the world's population and the national debt.

Scientific notation: To write a number in scientific notation, write it as a product of a number between 1 and 10 and a power of 10, in other words, as $\mathbf{N \times 10^{k}}$, where $1 \le N < 10$ and k is an integer.

Activity 3.12

a) Compute

i) 10^3 ii) 10^4 iii) 10^5

b) Based on the patterns you see in part a, how many digits would there be in 10^{100}?

c) Generalizing, how many digits in 10^n, where n is a positive integer?

d) Write the following in the form 10^n.

i) 1,000,000   ii) 10,000,000,000

e) Write 4,000,000 in scientific notation.

Solution: $4,000,000 = 4 \times 1,000,000 = 4 \times 10^7$

f) Write 50,000,000,000 in scientific notation.

Example 3-5.1: Write the following in scientific notation:

    a)   15,000,000
    b)   2,810,000

Solution: a) $15,000,000 = 1.5 \cdot 10,000,000 = 1.5 \times 10^7$

             b) $2,810,000 = 2.81 \cdot 1,000,000 = 2.81 \times 10^6$

Activity 3.13

a)   Based on Example 3-5.1, find a rule to convert the following to scientific notation.

    i)   4,500,000,000   ii)   50,800,000,000,000

b)   State your rule in words.

c)   Write 2.8×10^5 in standard notation.

d)   Write 4.76×10^8 in standard notation.

Now let's see how to write small numbers in scientific notation. A hydrogen atom weighs about 0.00000000000000000000000017 gram. To see how to write a small number like this in scientific notation, study the next example looking for any patterns.

Example 3-5.2: The following small numbers are written in scientific notation.

a) $0.05 = 5 \times 10^{-2}$ b) $0.051 = 5.1 \times 10^{-2}$

c) $0.002 = 2 \times 10^{-3}$ d) $0.4 = 4 \times 10^{-1}$

e) $0.00017 = 1.7 \times 10^{-4}$

Activity 3.14

a)   Write the following in scientific notation.

    i)   0.0003   ii)   0.000102

    iii)   0.00000034   iv)   0.0000000002003

b)   Write the following in decimal notation.

    i) 7.6×10^{-5} ii) 4.03×10^{-10}

To enter 1.24×10^{3} into Excel, you would enter the following:

You have to be careful when you look at the output in Excel. For example, if your output was **5.26E-10** you may quickly look at the output and think your answer was 5.26. The answer is actually 5.25×10^{-10}.

Activity 3.15

Write the following Excel output in decimal notation and in the form N x 10^{k}.

	A
1	2.30E+12
2	9.45E-05
3	2.30E+09
4	2.00E-10

If you want all of your output to appear in scientific notation, select the Home tab and go to general and then select Scientific from the menu as shown in the following figure.

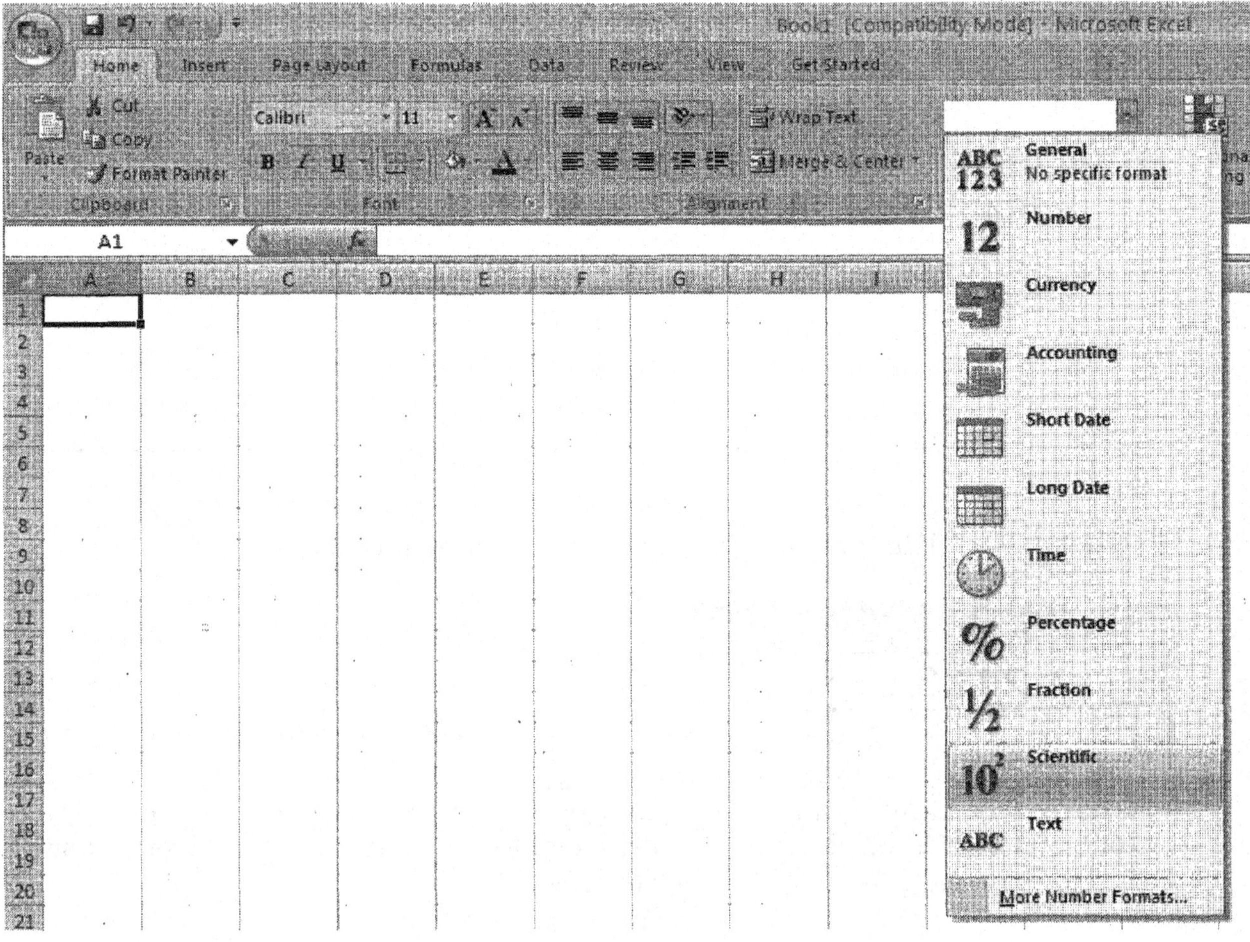

Homework Section 3-5

1. Write the following in scientific notation.

    a) 6,300,000 b) 0.00003 c) 0.000000203

    d) 7,904,000,000,000

2. Write the following in decimal notation.

    a) 4.5×10^{-7} b) 2.05×10^{9} c) 9.99×10^{-10}

3. Use Excel to compute $\dfrac{\left(5.6 \times 10^{15}\right)\left(3.17 \times 10^{-5}\right)}{7.3 \times 10^{9}}$. Round your answer to 3 decimal places.

<u>***Section 3-6: Exponential Growth***</u>

A theme that has run throughout American history as well as world history has been growth. As Donella Meadows states in her book *Beyond the Limits*,

> "Most societies, rich and poor, seek some kind of expansion as a remedy for their most immediate and important problems. In the rich world economic growth is believed to be necessary for employment, social mobility, and technical advance. In the poor world economic growth seems the only way out of poverty. And a poor family sees that many children can be a source not only of joy, but also of hope for economic security. Until other solutions are found for the legitimate problems of the world, people will cling to the idea that growth is the key to a better future, and they will do all they can to produce more growth."

Quantities can grow in two basic ways: **linearly** and **exponentially**. We will study both of these in detail in later chapters, but right now we want to just get a basic understanding of the difference between them. Quantities grow **linearly** when they grow by a fixed amount each time period and quantities grow **exponentially** when they grow by multiplication by a fixed fraction or percentage of their size in any period of time.

Suppose you invest $1,000 in a company that offers to pay you 10% interest per year. If you invest your money and let interest accumulate for 5 years, the money you accumulate may be as little as $1500 or as much as $1,610.51 depending if the money grew **linearly** or **exponentially**. Linear growth is equivalent to "simple interest" and exponential growth is equivalent to "compound interest".

Example 3-6.1:   Suppose you invest $1,000 in a company that pays you 10% interest per year, how much would you have at the end of 5 years if

      a) interest is calculated by simple interest,

      b) interest is compounded annually.

Solution: To calculate simple interest, take 10% of the principle, in this case $1,000, and this amount is added each period. Here our period is yearly. When interest is added to the principal at stated periods the interest is said to be *compounded*. The calculations are shown below.

Year	Simple Interest (Linear Growth)	Compound Interest (Exponential Growth)
1	1100	1000+.1(1000)=1100
2	1200	1100+.1(1100) = 1210
3	1300	1210+.1(1210) = 1331
4	1400	1331+.1(1331) = 1464.10
5	1500	1464.10+.1(1464.10) = 1610.51

Notice in simple interest your money is growing by a fixed amount ($100) each year, whereas when your money is compounded you get 10% of the previous year. The following Excel worksheet will calculate the amount of money in the account when the principal is compounded annually.

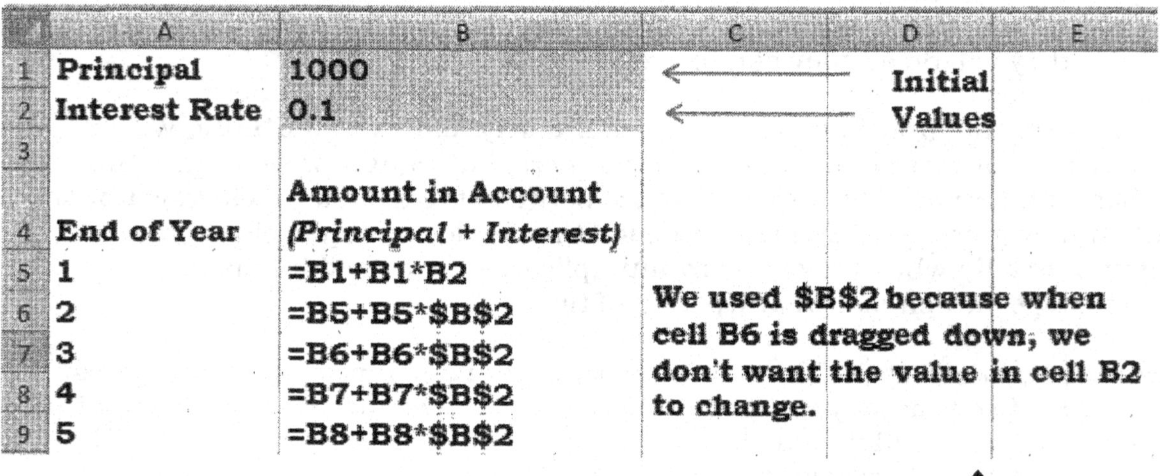

Using the last example, let's finish constructing the table below. You want to try and recognize the pattern and then express it in general form in the nth row.

<table>
<tr><td>Activity 3.16

Complete the table below.</td></tr>
</table>

End of Year	Amount in Account (Principal + Interest)	Amount in Account Factored Using Distributive Law Then Simplified
1	$1000 + .1(1000)$	$1000 + .1(1000) = 1000(1 + .1) = 1000(1.1) = 1{,}100$
2	$1000(1.1) + .1(1000)(1.1)$	$1000(1.1)(1 + .1) = 1000(1.1)^2 = 1{,}210$
3	$1000(1.1)^2 + .1(1000(1.1)^2)$	$1000(1.1)^2(1 + .1) = 1000(1.1)^3$
4		
5		
$\vdots$		
n		

Based on the last activity, we have the following generalized formula:

$$FV = PV(1 + r)^n \text{, where}$$

FV = future value
PV = present value
r = interest rate
n = number of years

How would your money grow if it is compounded monthly? The annual interest rate r would be divided by 12, the number of compounding periods in one year, and the number of years n is multiplied by 12 to get the total number of compounding periods. The formula would be

$$FV = PV\left(1 + \frac{r}{12}\right)^{(n \times 12)}.$$

To generalize this formula further, if the compounding period is k times a year we have

$$FV = PV\left(1 + \frac{r}{k}\right)^{(n \times k)}.$$

Example 3-6.2: Calculate the future value of $1,000 earning an annual interest rate of 6% after 10 years, if compounded:

a)   quarterly
b)   daily

Solution: a) $FV = 1000\left(1 + \dfrac{0.06}{4}\right)^{(4 \times 10)} = \$1,814$

b) $FV = 1000\left(1 + \dfrac{0.06}{365}\right)^{(365 \times 10)} = \$1,822$

You can use the following Excel worksheet to do the calculations.

	A	B
1	PV =	1000
2	r =	0.06
3	k =	365
4	n =	10
5		
6	FV =	=B1*(1 + B2/B3)^(B3*B4)

Another example of exponential growth is population growth.

Activity 3.17

Suppose the population of a certain city has been growing steadily and has been doubling every 20 years.

a) If the population in 1990 was 5,000, complete the following.

Year	Population
1990	_____
2010	_____
2030	_____
2050	_____

b) If **t = the number of 20 year time periods after 1990**, a formula for the population in terms of t is

$$P = 5000(2)^t.$$

What would be the value of t if you wanted the population in 2010? 1990? 1970?

c) Use the formula above and Excel to find the population in 2050. Your answer should match the answer you have in part a.

Unit 3 Review Exercises

1. Use Excel to determine how much money you would have at the end of 1 year (n = 1) if you invested $1000 at an interest rate of 12% (r = 0.12) compounded monthly (k = 12). Round your answer to the nearest cent.

2. Suppose you invested $ 3000 in a bank that advertised an interest rate of 3.75% compounded quarterly, how much would you have at the end of 3 years? Round your answer to the nearest cent.

3. Use Excel to evaluate the given expressions. Round all answers to <u>three</u> decimal places. It would be helpful to write out your keystrokes before entering anything into Excel.

    a) $$\dfrac{5.21(-1.54)}{(^-2.75)^5 + 2.45}$$

    b) $1.32(^- 5.76) - (^- 7.21)^3 + 0.542^2 / (2.67 - 1.01^2)$

4. Evaluate $\dfrac{x(1000 - 20x^2 + x^3)}{1850}$ when x = 4.72. Round the answer to three decimal places.

5. a) Suppose you invested $1500 in a bank that advertised an interest rate of 4% compounded monthly, how much would you have at the end of 2 years? What if the interest rate was compounded daily?

    b) If $10,000 is invested at an interest rate of 3% per year, compounded semiannually, find the value of the investment after 5 years. After 10 years.

6. Calculate the following.

    a) The time in seconds it takes a computer to check n memory cells is found by evaluating $(n/2650)^2$. Find the time to check 1000 cells.

    b) The population of China, in billions, can be approximated by evaluating $1.15(1.014)^t$, where t is the number of years since the start of 1993. According to the formula, what was the approximate population of China at the start of 2000? 2006?

---

7.   Write the following in scientific notation:

   a)   0.00000045   b)   1,200,000,000,000
   c)   0.0000000000298

8.   Write the following in decimal notation:

   a) 4.6×10^{-7} b) 5.07×10^{9}

9.   Use Excel to compute the following. Write your answer in scientific notation and round to two decimal places.

$$\frac{\left(4 \times 10^{-6}\right)\left(5.8 \times 10^{12}\right)}{3.5 \times 10^{4}}$$

10.   Use the Rules of Exponents to simplify the following. Write all answers with positive exponents.

   a) $\left(2x^{6}y^{-1}\right)^{5}$ b) $\frac{x^{3} \cdot x^{-5}}{x^{6}}$ c) $\left(\frac{xy^{7}}{y^{3}}\right)^{-2}$

11.   Rewrite 16^{t} so the base is 2.

12.   You know that $2^{1001} \cdot 2^{n} = 2^{2000}$. What is n?

13.   You know that $\frac{2^{1000}}{2^{n}} = \frac{1}{16}$. What is n?

14.   So far we have defined a^{n} for any integer n and for any a. This is not the end of the story though, because n is not limited to the integers. What might $4^{1/2}$ be? Motivate your answer as well as you can.

UNIT 4
Functions

Section 4-1: Introduction to Functions

Functions are one of the most fundamental concepts of mathematics. We will see that they are useful for describing (**modeling**) many real-world problems. Let's start by looking at some examples of functions.

In Unit 1 we looked at the following sequence of figures made with matches.

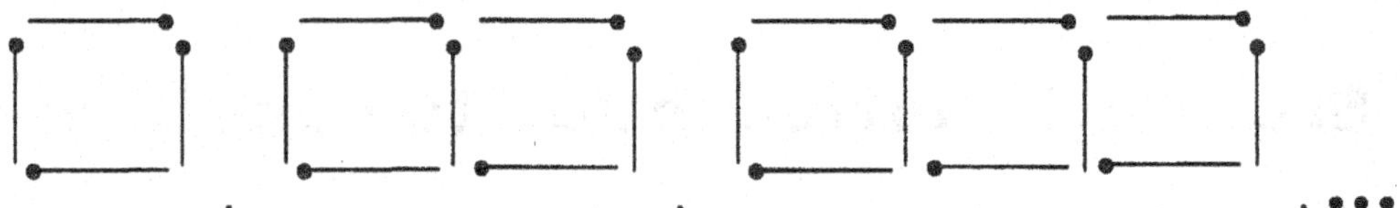

The number of matches used is a **function** of the number of squares. The "number of squares" is called the **independent variable**, or the **input**. The "number of matches" is called the **dependent variable**, or the **output**, since the number of matches _depends_ on the number of squares.

Another example we looked at was the value of a car. Suppose a new car depreciates 20% each year. We can then say that the value of a car is a function of the number of years it is owned. In this case the "number of years a car is owned" is the independent variable, and the "value of the car" is the dependent variable.

> **Definition**: A **function** is a rule that, for each valid input, assigns _one and only one_ output.

This definition can be visualized with the following diagram. You have an input, a rule, and an output.

$$\text{input} \rightarrow \boxed{\text{rule}} \rightarrow \text{output}$$

Suppose you purchased a new car for \$20,000 and the car depreciates 20% each year. Here if we input the number of years you own the car, we will get output the value of the car. The rule that assigns the output is $20{,}000(0.8)^n$.

$$\text{input} \rightarrow \boxed{20{,}000(0.8)^n} \rightarrow \text{output}$$

The set of all possible inputs is called the **domain** of the function and the set that consists of all possible outputs is called the **range** of the function.

Depending on the situation, a function can be best represented as a **formula (symbolically)**, **data in a table (numerically)**, **or as points on a graph (graphically)**.

Example 4-2.1: Suppose you purchased a new car for $20,000 and the car depreciates 20% each year. We can get the value of the car by the formula

$$V = 20{,}000(0.8)^n \,,$$

where n is the number of years you own the car. Here we are using a formula to represent the function. We could also put the data in an input-output table.

Number of years car owned n (input)	1	2	3	4	5	...
Value of car V (output)	16,000	12,800	10,240	8,192	6,553.60	...

We can also write the data in the table as **ordered pairs**. The ordered pairs in this example would be (1, 16000), (2, 12800), (3, 10240), (4, 8192), and (5, 6553.60). The first component in the pair is the input variable and the second component is the output variable. These ordered pairs are plotted below. Data represented this way is said to be plotted on the **Cartesian plane**.

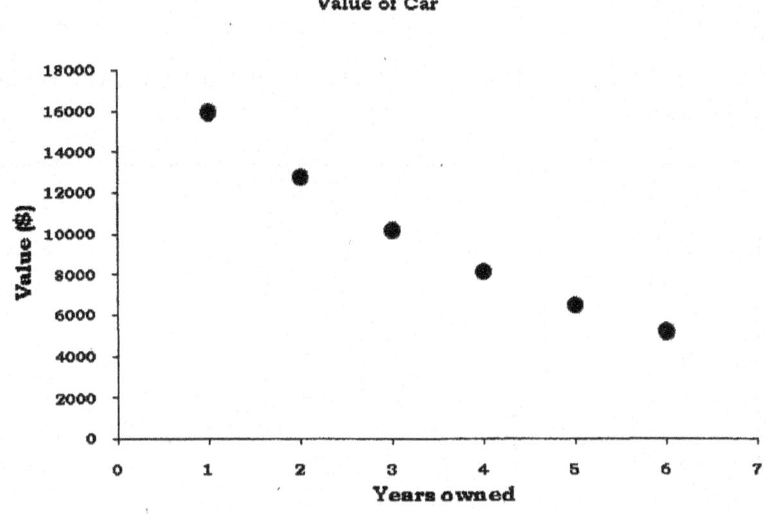

The following graph does not represent a function because there is an input, 3, which gives two different outputs. When you input 3 you get an output of 1 and 5. Remember to be a function for each input you can get one and only one output.

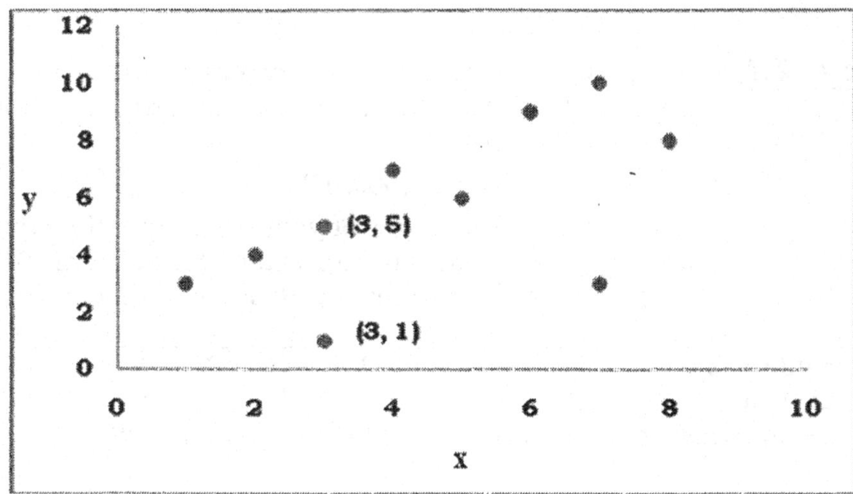

It may be helpful to think of the vertical line test when determining if a graph is a function.

Vertical Line Test
A curve is the graph of a function if and only if no vertical line intersects the curve more than once.

You can see that the last graph does not represent a function because a vertical line intersects the points (3, 5) and (3, 1).

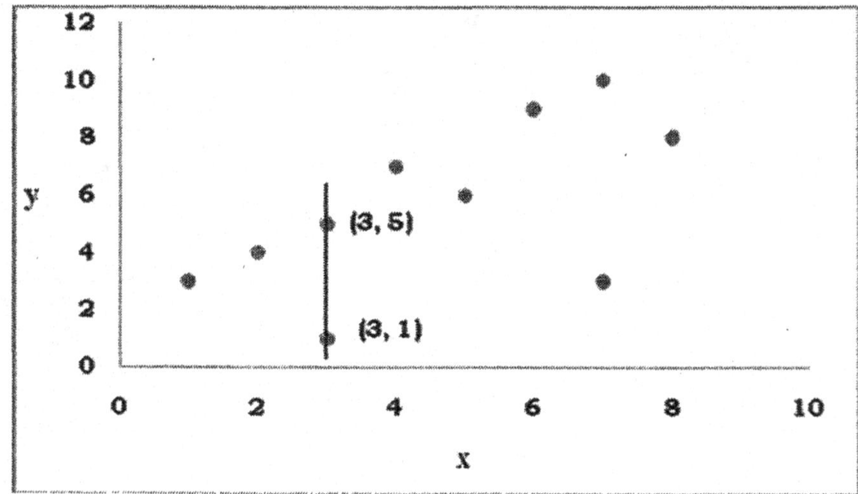

: Does the following graph represent a function?

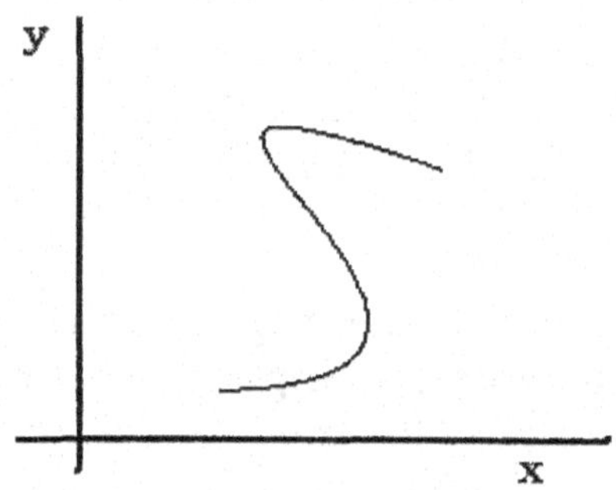

Solution: No, since a vertical line intersects the graph in more than one point.

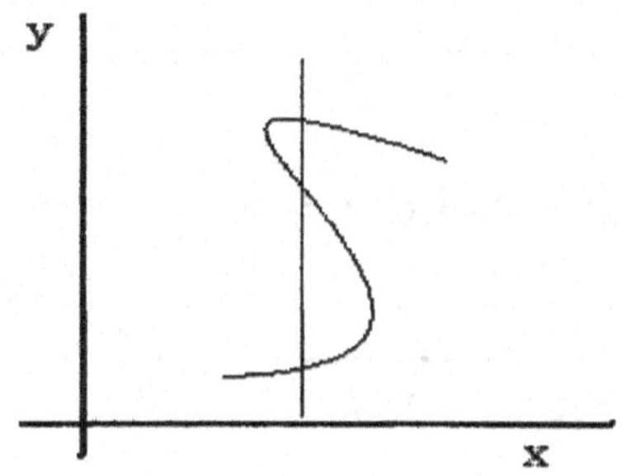

Activity 4.1

Why does $V = 20{,}000(0.8)^n$ represent a function?

Hint: Go back to the definition.

Note: When graphing a function the input variable (**independent variable**) is placed on the horizontal axis and the output variable (**dependent variable**) is placed on the vertical axis. This is illustrated in the figure below.

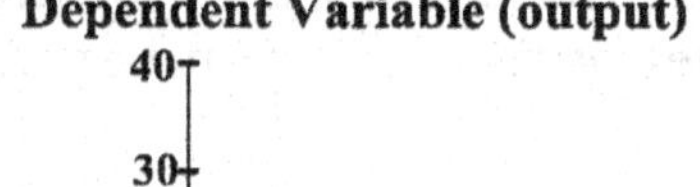
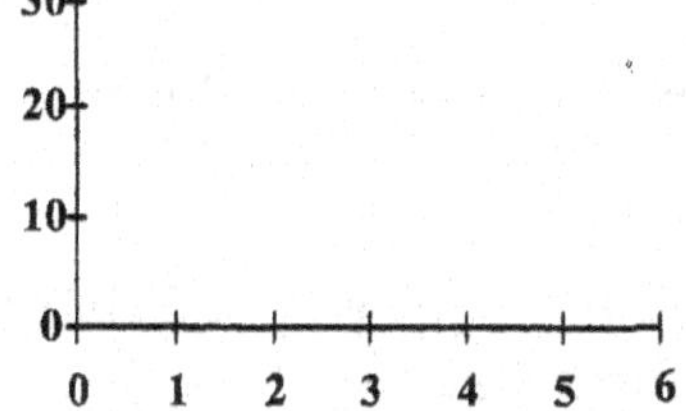

Example 4-2.3 : Is y a function of x in the following **relation** (set of ordered pairs)?

x (input)	2	2	4	4	5
y (output)	3	1	5	3	7

Solution: For a relation to be a function, each input has to have a unique output. In this relation when a 2 is input the output is 3 and 1. Since the output is **not** unique for the input 2, this relation is **not a function.**

♦

Activity 4.2

Is y a function of x in the following relation? Explain.

x (input)	3	4	5	7	10
y (output)	10	10	10	10	10

Activity 4.3

a)   Not all functions have to mathematical, which of the following represent functions? If it does not represent a function, state why not.

  i)   If you input a person's name, you get output the number of letters in their name.
  ii)   If you input a person, you get output the make of their car.
  iii)   If you input a person, you get output their social security number.
  iv)   If you input a set, you get output the number of elements in the set.

b)   Functions are introduced in about the third grade with problems like the following:

Use the table to help find each secret rule. Finish the table and write the rule in words and as a formula.

(continued on next page)

---

i)

In	3	4	5	7	10
Out	7	8	9	__	__

Rule: _______________________________

ii)

In	5	6	8	10	15
Out	14	15	17	__	__

Rule: _______________________________

c) We can take the rule we find and write it as a formula for the function. Given the following input-output table, fill in the rule for the function.

x (input)	2	9	14	17
y (output)	0	7	12	15

x (input) $\longrightarrow$ [ ] $\longrightarrow$ y (output)

Write the function as a formula.

y = _______________________________

Activity 4.4

It is important to realize that graphs are not just pictures, but are abstract representations of relationships. The figure on the left shows the path of a golf ball that has been hit. Sketch a rough graph that shows how the speed of the ball changes.

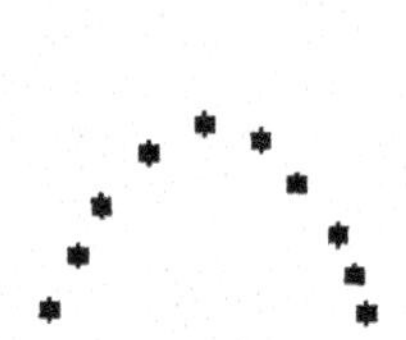

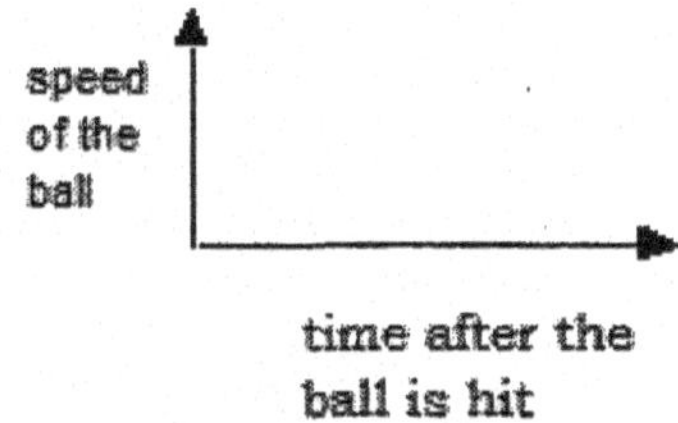

Activity 4.5

For each of the graphs below give a table of values that could possibly go with
the graphs.

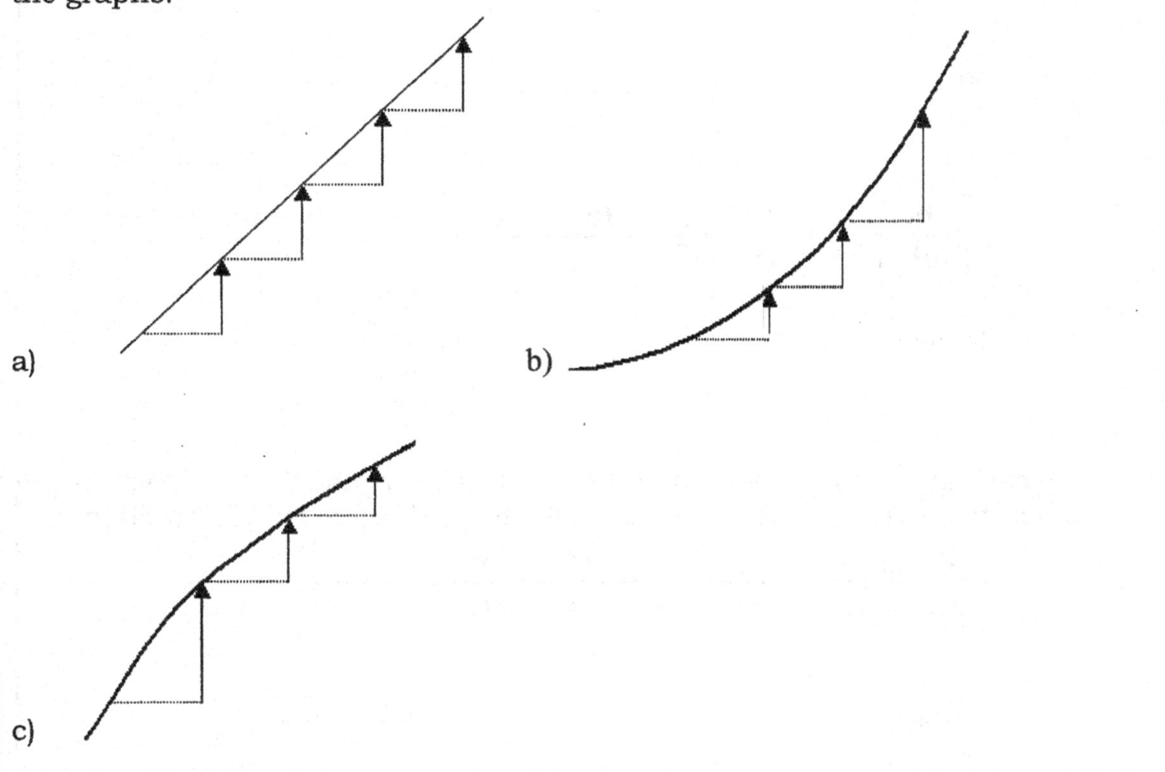

The following table gives the maximum daily drug dosage of ampicillin for children who weigh less than 10 kilograms (about 22 pounds).

Daily Ampicillin Dosage by Weight	
W	D
Child's Weight	Daily Maximum Dosage
0	0
1	50
2	100
3	150
4	200
5	250
6	300
7	350
8	400
9	450
10	500

Source: Math for Meds: Dosages and Solutions, 6th ed. (San Diego: W.I. Publications, 1990): p. 198.

Weight and dosage are called **quantitative variables**. We say that

"dosage depends on weight"   or   "dosage is a function of weight."

Because dosage depends on weight, we can think of weight as the **input variable** and dosage as the **output variable**.

According to the table, the dosage for a child weighing 4 kilograms is 200 milligrams of ampicillin. Mathematicians use a shorthand notation for this statement and write it as, D(4) = 200. (Read D(4) = 200 as "**D of 4 equals 200**.") This notation indicates a relation in which one variable is a function of another. When you see D(4) = 200, you can also think of this as when you input a weight of 4 you get a dosage of 200 milligrams for output. You could also write this as the ordered pair (4, 200).

Activity 4.6

a)   Using the Daily Ampicillin Dosage by Weight table above, find the following:

   i)   D(9) = _____   ii)   D(1) = _____

b)   The maximum safe dosage for a child weighing 10 kg is 500 mg. Write this statement using function notation.

The following may help give more meaning to function notation.

$$D\,(4) \;=\; 200$$

Name of · input · output
function

Example 4-3.1: A function is defined by the formula $f(x) = 4x - 1$. Find the
following.

    a) $f(1)$ b) $f(3)$

Solution: The function $f(x) = 4x - 1$ can be viewed as

$$x \text{ (input)} \rightarrow \boxed{\begin{array}{c}\text{multiply by 4 then}\\ \text{subtract 1 from the result}\end{array}} \rightarrow \text{output}$$

a) To find $f(1)$ we input 1, multiply by 4 to get 4 and then
subtract 1 to get 3. So, $f(1) = 3$.

A more compact way to write this is
$$\boxed{f(1) = 4 \cdot 1 - 1 = 3}\;.$$

Everyplace we had an x in $4x - 1$; we replaced it with a 1.

b) $\boxed{f(3) = 4 \cdot 3 - 1 = 12 - 1 = 11}$

◆

Activity 4.7

Given the following function rules, find f(3) and f(5).

    a) $f(x) = x + 9$

         $f(3) = $ ______________________

         $f(5) = $ ______________________

    b) $f(x) = x^2 - 4x + 7$

         $f(3) = $ ______________________

         $f(5) = $ ______________________

Example 4-3.2: Fifty-five percent of public 4-year college students graduate with debt. The following table gives the average debt of these students from 2001 to 2007. The amount of debt has been adjusted for inflation.

Year of Graduation	2001	2002	2003	2004	2005	2006	2007
Amount Of Debt	17,400	17,200	17,900	18,000	18,200	18,800	18,800

Source: www.collegeboard.com/html/costs/aid/4_1_loans.html

The figure below is a graph of the data with the straight line representing an approximation of the function. Finding this approximation will be covered in some later units. Right now just realize that finding an approximation to data is important in making predictions from a set of data.

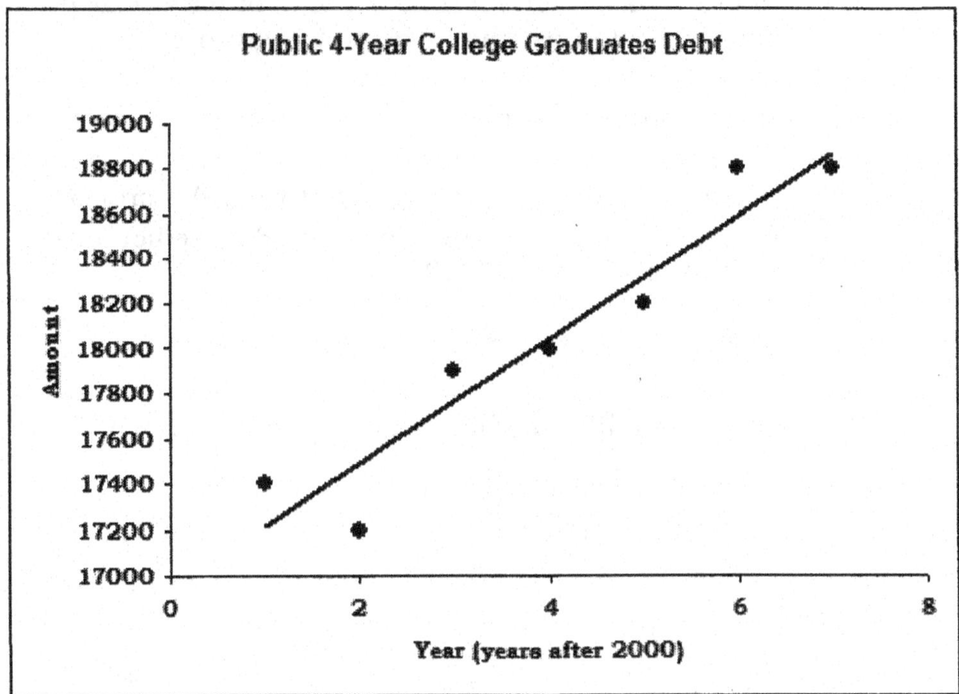

In order to predict what might happen in the future, a formula will be useful. The linear approximation graphed above has the formula

$$A(t) = 275t + 17{,}218$$

where *A(t)* is the approximate amount of college debt and *t* is the time in number of years since 2000 that the student graduated.

Using the formula we can estimate the amount of debt for a student who will graduate in 2008. The estimated amount of debt for a student who graduates with a debt would be

$$A(8) = 275(8) + 17{,}218 \approx \$19{,}418.$$

Activity 4.8

For the function $A(t) = 275t + 17{,}218$ that models college graduate debt, explain what $A(0) = \$17{,}218$ means in the given context.

Example 4-3.3: We can use different functions as models of the same data. Another model for the amount of debt for 4-year public college graduates who have debt is given by

$$A_1(t) = 17{,}218.71 \cdot 10^{0.0066t}$$

where A_1 is used to distinguish it from the previous function. What does the second model predict for the amount of debt in 2008?

Solution: $A_1(8) = 17218.71 \cdot 10^{0.0066 \cdot 8} \approx \$19{,}444.67$

This estimate is a little higher than our prediction with $A(t)$.

♦

Homework Sections 4-1 – 4-3

1. Complete the table.

   a) $f(x)=2x^2+1$

x	$f(x)$
-1	
0	
1	
2	
3	

   b) $g(x)=\left|3x-5\right|$

   Note: $|x|$ (read "the absolute value of x) is defined as follows:

   $$x=\begin{cases} x, & \text{if } x \geq 0 \\ -x, & \text{if } x < 0 \end{cases}$$

x	$g(x)$
-2	
-1	
0	
1	
2	

2. Evaluate the function at the indicated values.

   a) $g(x)=\dfrac{1-x}{1+x}$

   i) g(-2)   ii) $g\left(\dfrac{1}{2}\right)$ iii) g(a)

   b) $h(t)=t+\dfrac{1}{t}$

   i) $h(-1)$ ii) $h(x)$ iii) $h(x-1)$

3.   What is the domain of the following functions?

a) $f(x) = x^3 + 1$ b) $g(x) = \dfrac{1}{x-4}$

4.   The graph below shows the energy use per capita for the United States and the world between 1960 and 2005.

   a)   What was the energy use for the United States in 1980?
   b)   When was the energy use in the United States the highest?
   c)   How would you describe the energy use in the world?
   d)   Between what years did the energy use in the United States decrease?
   e)   Is the energy use in the United States a function of year? Explain.

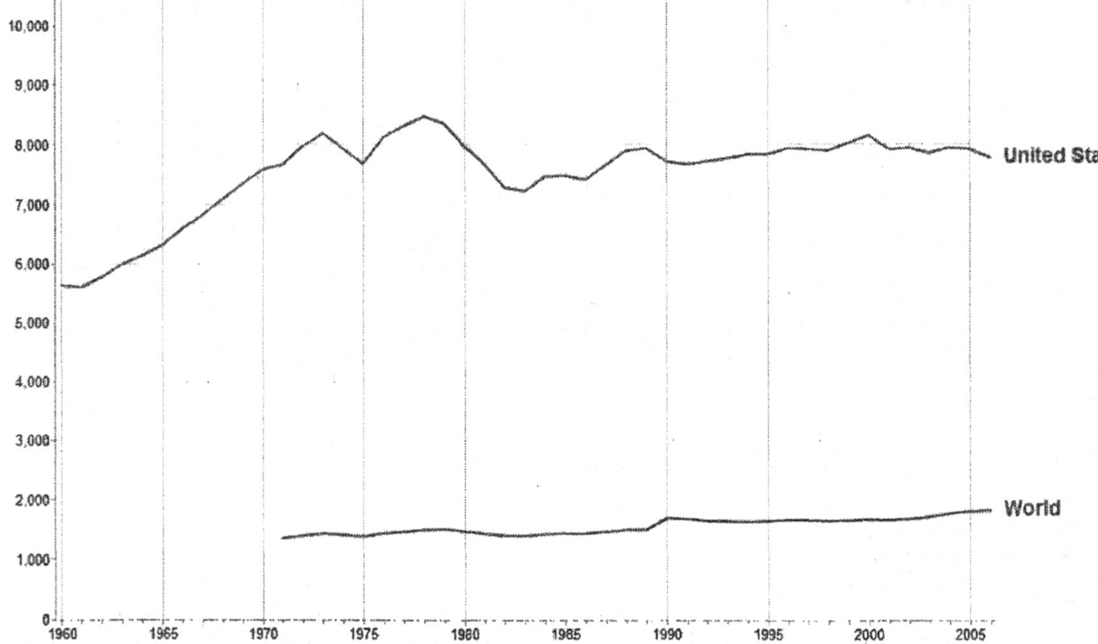

Section 4-4:   Graphing Functions with Excel

In the last section we looked at two functions

$$A_1(t) = 17{,}218.71 \cdot 10^{0.0066t} \ \text{ and}$$

$$A(t) = 275t + 17{,}218 \,.$$

Let's use Excel to graph these functions so that we can compare them. We will look at the functions from t = 0 to 20.

1. Enter the worksheet as it appears below in cells A1 to C22, using the formulas for Cell B2 and Cell C2 given to the right in the spreadsheet. Highlight the columns of variables you want to plot.

2. Click on the tab labeled **Insert,** and then select **Scatter** from the **Charts** menu.

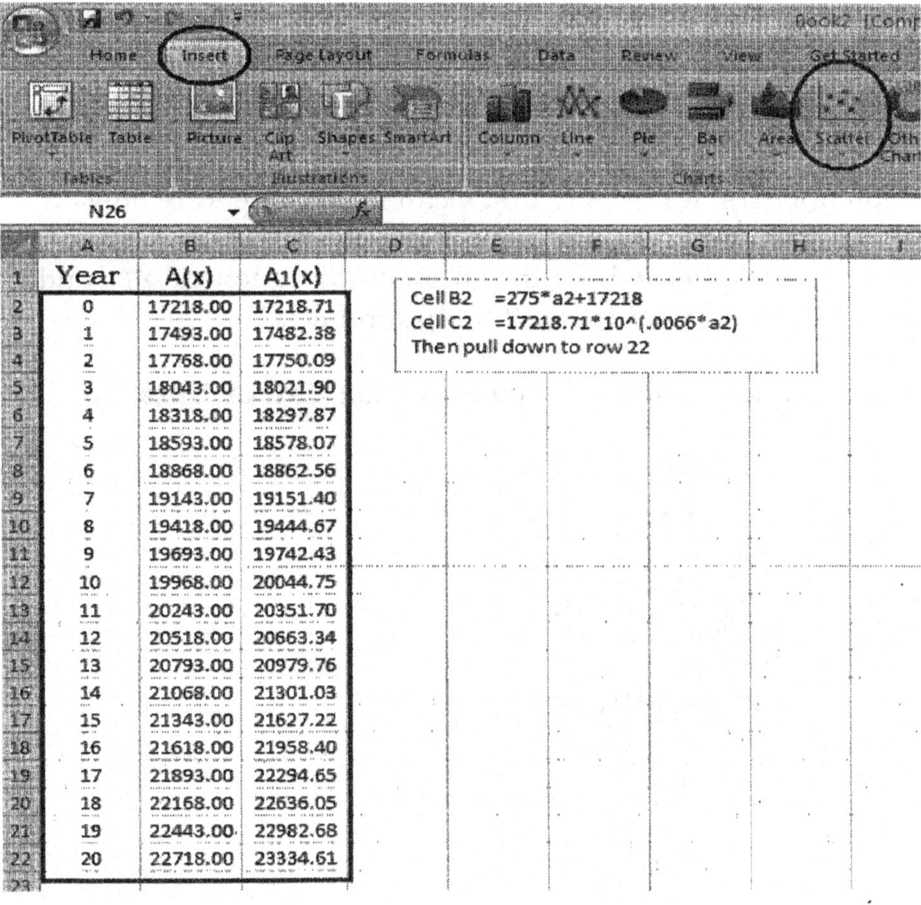

Year	A(x)	A1(x)
0	17218.00	17218.71
1	17493.00	17482.38
2	17768.00	17750.09
3	18043.00	18021.90
4	18318.00	18297.87
5	18593.00	18578.07
6	18868.00	18862.56
7	19143.00	19151.40
8	19418.00	19444.67
9	19693.00	19742.43
10	19968.00	20044.75
11	20243.00	20351.70
12	20518.00	20663.34
13	20793.00	20979.76
14	21068.00	21301.03
15	21343.00	21627.22
16	21618.00	21958.40
17	21893.00	22294.65
18	22168.00	22636.05
19	22443.00	22982.68
20	22718.00	23334.61

3. Choose the second plot style in the left column from the drop-down menu.

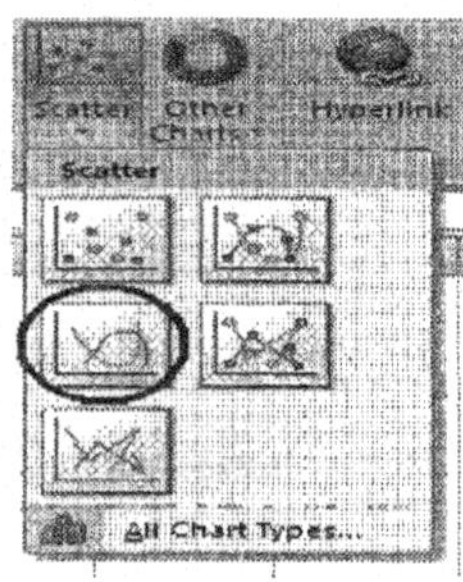

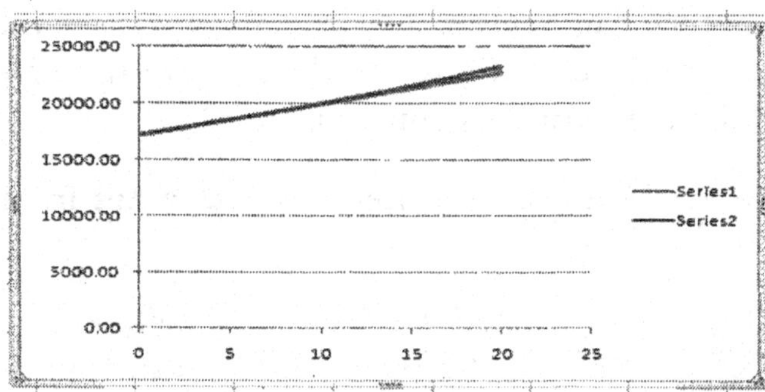

4. Once the graph appears, we need to first clean it up. If we were only graphing one function we could first remove the legend by clicking on it and pressing **Delete**. Since we have the graph of two functions $A(x)$ and $A_1(x)$ let's change Series 1 and Series 2 to $A(x)$ and $A_1(x)$.

Right click anywhere in the plot area and click **Select Data**.

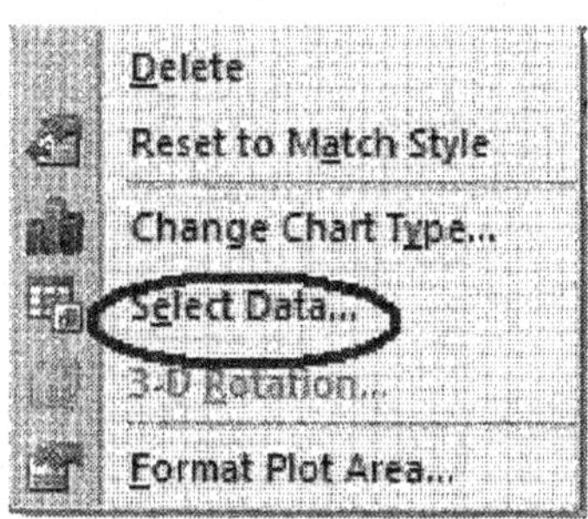

5. You will notice that Series 1 is highlighted, click **Edit**.

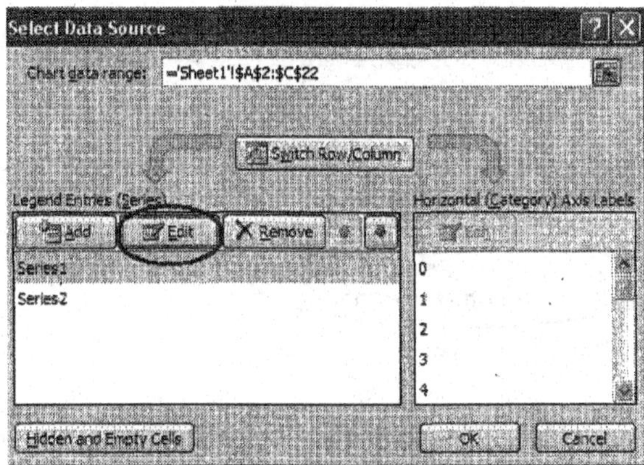

6. Type A(x) in the blank box and then click **OK**.

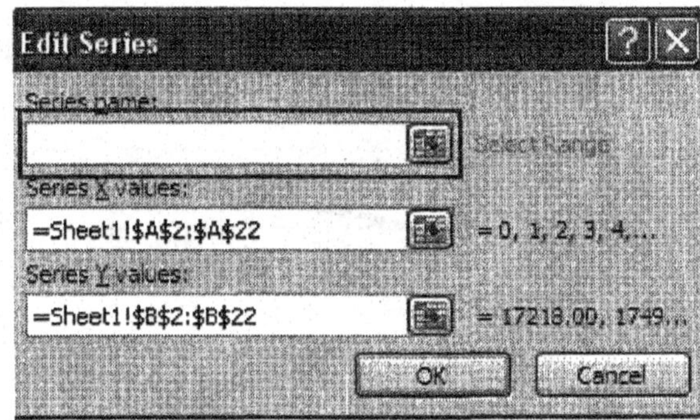

7. Highlight Series 2, click edit, type A1(x) in the blank box, and click OK.

8. Next, add a title and axis titles by clicking anywhere on the scatter plot, selecting the **Layout** tab at the top and then clicking on **Chart Title**.

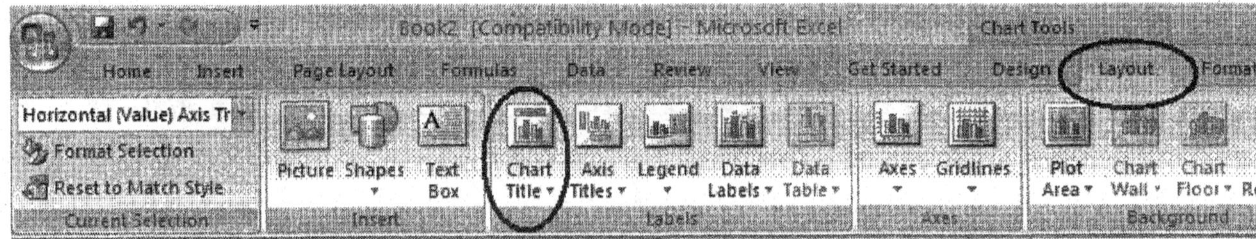

9. Click **Above Chart**, and type in a title (Graph of A(x) and A1(x)).

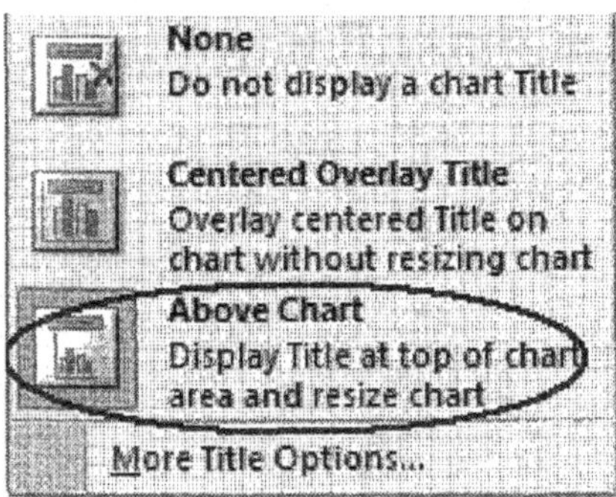

10. Now click on **Axis Titles**. For the horizontal title, you want it to appear below the x-axis. Type in *t (years since 2000)*. For the vertical title, you want the **Rotated Title** option. Type in *Amount of Debt*.

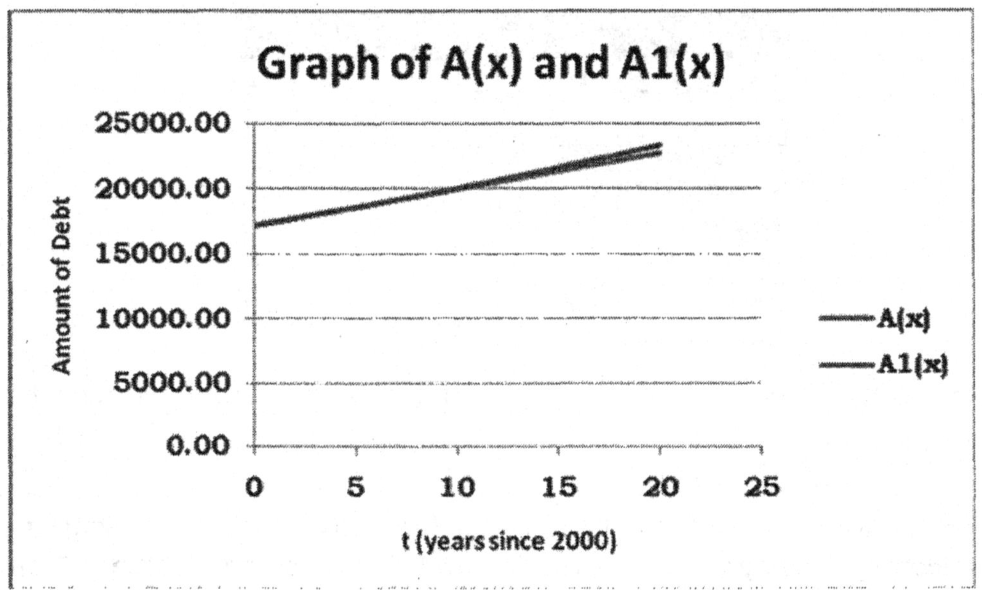

You can see from the graphs of the two functions that for values of t from 0 to 15 the functions are almost identical, but after that the functions begin to separate indicating that $A_1(x)$ will give higher approximations than *A(x)*.

Use Excel to graph the function $f(x) = 4x^5 - 5x^4 - 40x^3$ from x = -3 to 4.
Your graph should appear like the one below.

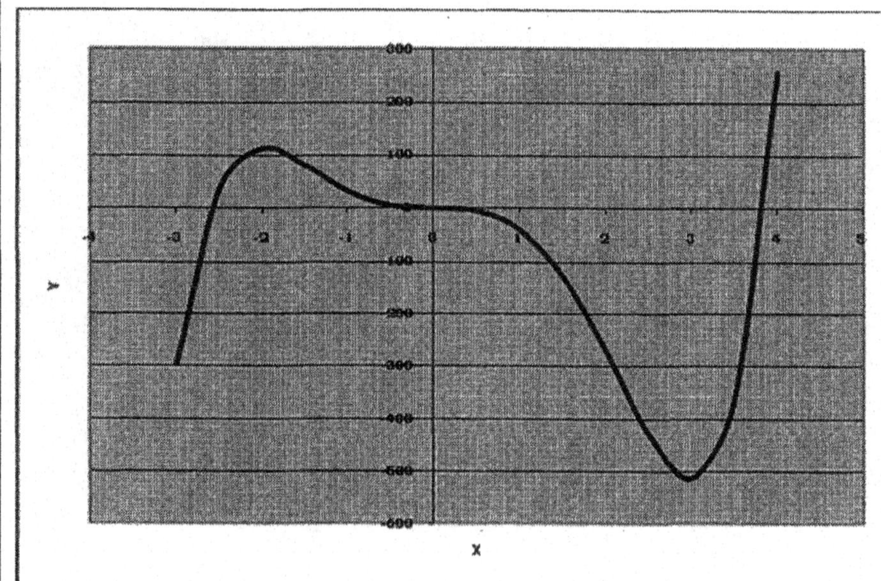

Homework Section 4-4

1.   Use Excel to graph the following functions.

   a) $f(x) = \dfrac{x+3}{2}$

   b) $g(x) = x^2 - 4$

   c) $h(x) = 4x^2 - x^4$

   d) $G(x) = |2x - 2|$

   Note: Type = **abs(number)** to find the absolute value of a number.

2.   Graph the function $f(x) = (x-c)^3$ for the following values of c:

$$0, 2, 4, 6$$

How does the value of c affect the graph?

You have probably heard your parents or grandparents say something like, "I remember when gasoline was 35 cents a gallon!" As time goes by most items become more expensive due to inflation. In the past a dollar bought more than it does today, but you have to remember that most people earned less than they do now. In order to compare costs of items across years we use the Consumer Price Index (CPI). The CPI is compiled by the Bureau of Labor Statistics (BLS) monthly and is based upon a 1984 base of 100. The table below gives the CPI for some selected years. A CPI of 185 indicates 85% inflation since 1984. Since this is an index it only tells you the total inflation since the base year. A CPI of 9.9 in 1913 indicates that prices multiplied over ten times from 1913 to 1984. It has doubled again from 1984 to 2006 when the CPI topped 200.

Year	1913	1923	1933	1943	1953	1963	1982	1983	1993	2006	2008
CPI	9.9	17.1	13	17.3	26.7	30.6	96.5	99.6	144.5	201.6	215.3

The CPI is measured by pricing the items on a list representative of a typical urban household budget and calculating the average change in prices paid by urban consumers for the items on the list over a period of time. The BLS classifies the items into the following eight groups:

- FOOD AND BEVERAGES
- HOUSING
- APPAREL
- TRANSPORTATION
- MEDICAL CARE
- RECREATION
- EDUCATION AND COMMUNICATION
- OTHER GOODS AND SERVICES

Let's look at an example to help understand how the CPI is constructed. Assume that students at Keene State College only buy three items: books, music CD's and coffee. We want to construct a KPI (Keene Price Index) for the students. First, we would conduct a survey of spending habits in a base year, suppose it is 2000. Suppose the results are given in the table below.

<u>Item</u>	<u>Average Expenditure per Month in 2000</u>
Books	$60
Cd's	$32
Coffee	<u>$8</u>
Total	$100

Suppose the cost of the same items in 2009 is given below.

Item	Average Expenditure per Month in 2009
Books	$90
Cd's	$45
Coffee	$15
Total	$150

To calculate the KPI, based on 2000 = 100, is

$$KPI = \frac{\cos t \ of \ budget \ in \ 2009}{\cos t \ of \ budget \ in \ 2000} \times 100 = \frac{150}{100} \times 100 = 150$$

So, the KPI in 2009 is 150, which means that students' cost of living increased 50% over 9 years.

Information about the CPI can be obtained on the Bureau of Labor Statistics Web site. (*http://www.bls.gov/cpi/home.htm*)

Here is an example of how the CPI is used to compare costs or salaries in different years. Suppose on 1982 the starting salary for a particular job was $22,000 and was $46,000 in 2006. Who had more "buying power", the person starting the job in 1982 or the person in 2006? Remember the salary was less in 1982 but items were also cheaper in 1982. The CPI was 96.5 in 1982 and 201.6 in 2006, which means for every $96.50 you spent toward goods and services in 1982 you would have to spend $201.60 in 2006. This means that prices rose by a multiple of $\frac{201.6}{96.5} = 2.09$. So to have the same "buying power" the salary should also increase by a multiple of 2.09.

$$2.09 \cdot \$22,000 = \$45,980$$

So the two have about the same "buying power" meaning salaries did not increase from 1982 to 2006.

Example 4-5.1:   If an item costs \$1.00 in 1953, how much would it cost in 2008?

Solution: The CPI was 26.7 in 1953 and 215.3 in 2008, so prices rose by a multiple of $\dfrac{215.3}{26.7} = 8.06$, this number is called the inflation factor for comparing prices in 1953 to prices in 2008, prices increased by a factor of 8.06. So something that cost \$1.00 in 1953 would cost about \$8.06 in 2008.   ♦

Activity 4.10

If an item cost \$1.00 in 2008, what would it have cost in 1953?

Activity 4.11

The graph below shows the CPI as a function of time. How would you describe the function? Are there any trends that are apparent?

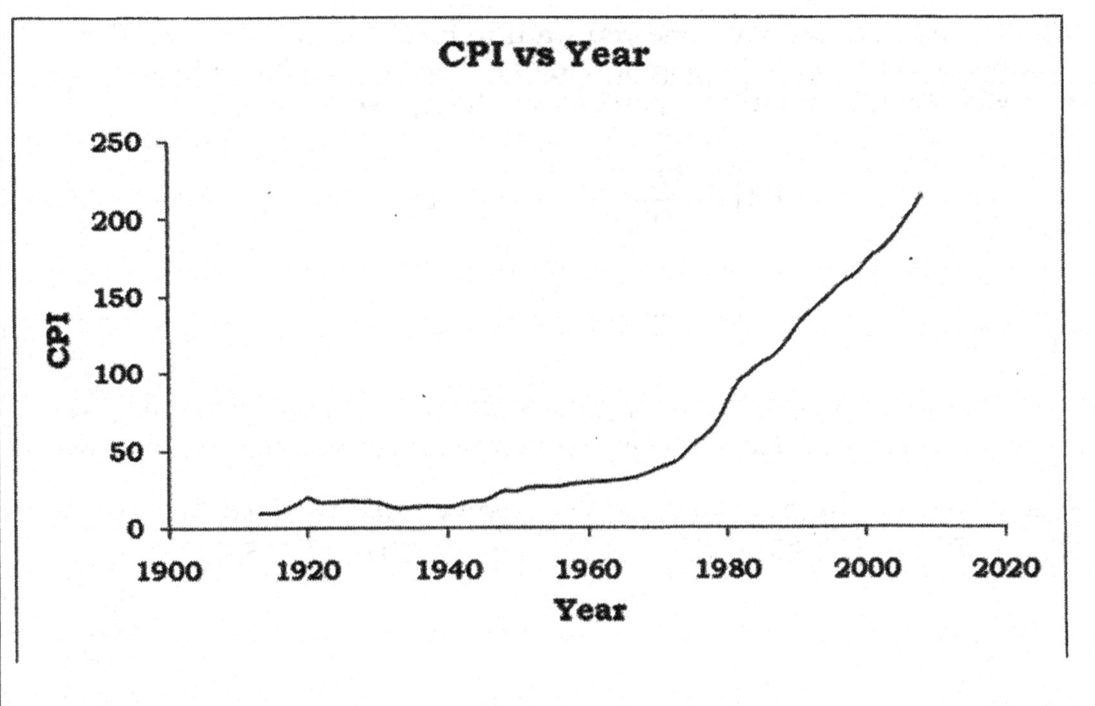

According to *The American Heritage® Dictionary of the English Language, Fourth Edition,* Copyright © 2000 Published by Houghton Mifflin Company **inflation** is defined as

> *a persistent increase in the level of consumer prices or a persistent decline in the purchasing power of money, caused by an increase in available currency and credit beyond the proportion of available goods and services.*

So if we want to know the inflation rate over the last 12 months (the commonly published inflation rate number) we would subtract last year's CPI from the Current CPI and divide by last year's number and multiply the result by 100 and add a % sign. The formula for calculating the Inflation Rate looks like this:

$$100\left(\frac{Current\ CPI - Old\ CPI}{Old\ CPI}\right)$$

So if exactly one year ago the Consumer Price Index was 178 and today the CPI is 185, then the calculations would be:

$$100\left(\frac{185-178}{178}\right) = 100\left(\frac{7}{178}\right) = 3.93\%.$$

So over the sample year inflation was 3.93%.

Note: This formula is used any time you want to find a percent change. For example suppose that today the population of Keene is 22,893 and in 2005 the population was 22,203, to find the percentage change we have

$$100\left(\frac{22893-22203}{22203}\right) = 3.11\%.$$

Activity 4.12

The table below gives the CPI from 2002 through 2008. Use Excel to calculate the inflation rate for those years and graph inflation rate as a function of time.

Year	2002	2003	2004	2005	2006	2007	2008
CPI	179.88	183.96	188.9	195.3	201.6	207.34	215.3

Unit 4 Review Exercises

1. Given $D(t) = 2.45t^3 + 0.1t - 120$, find the following.
    a)   D(5)   b)   D(4.2)

2. Using your calculator, find H(3.6) if $H(x) = 1.2x^4 + 0.5x - 6.4$.

3. Is y a function of x in the following relation? Explain.

x	2	4	6	7	9	10
y	12	15	15	16	17	19

4. Given $f(x) = 3x^2$. The domain of this function is all real numbers, what
    is the range? Remember the range is your output values.
    Hint: No matter what value you input for x, what can you say about the
    output?

5. Explain why the volume of a sphere is a function of the radius of a
    sphere. The formula for the volume of a sphere is $V = \frac{4}{3}\pi r^3$. What is
    the domain of the function?

6. Explain why the following graph does not represent a function.

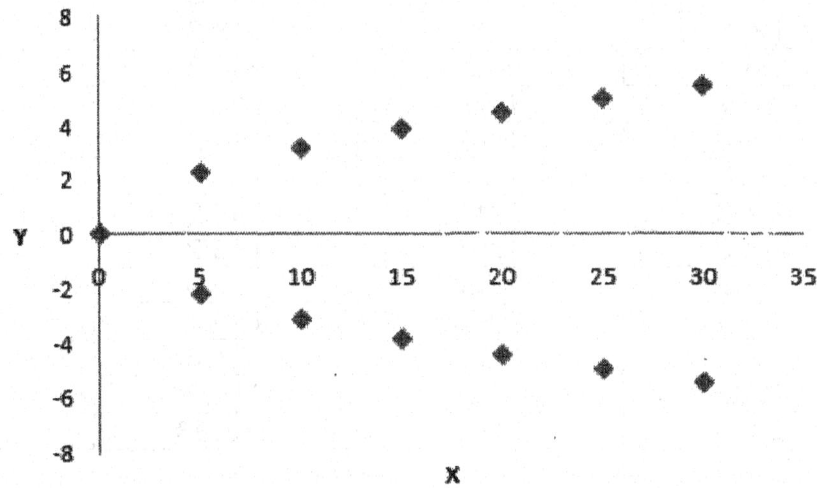

7. Use Excel to graph the function $V = \frac{4}{3}\pi r^3$.

8. Suppose an item cost $1.00 in 2008, how much would the same item have cost in 1982? Remember, this number is called the inflation factor.

Year	1913	1923	1933	1943	1953	1963	1982	1983	1993	2006	2008
CPI	9.9	17.1	13	17.3	26.7	30.6	96.5	99.6	144.5	201.6	215.3

9. Complete a table of inflation factors from 1913 to 2008.

Year	1913	1923	1933	1943	1953	1963	1982	1983	1993	2006	2008
Inflation factor											1.00

10. Using the data from the table in problem 9, graph the inflation factors with reference year 2008 on Excel. The independent variable will be year and the dependent variable will be inflation factors.

11. The terms current dollars or nominal dollars refer to an amount in actual dollars not adjusted for inflation. Use Excel to complete the following table:

Year	Nominal Price per gallon	CPI	Price in Real $ (2009)
1980	1.28	82.4	$1.28\left(\dfrac{216.33}{82.4}\right) = 3.36$
1982	1.42	96.5	
1984	1.22	103.9	
1986	.89	109.6	
1988	1.00	118.3	
1990	1.23	130.7	
1992	1.31	140.3	
1994	1.25	148.2	
1996	1.38	156.9	
1998	1.24	163.0	
2000	1.76	172.2	
2002	1.50	179.9	
2004	1.93	189.9	
2006	3.12	201.6	
2008	4.40	215.3	
2009	2.70	216.33	

12. Using the data from the table in problem 11, graph real price in 2009 dollars on the vertical axis and year on the horizontal axis on Excel. Wherever the graph of real prices is increasing, the nominal price is going up faster than inflation.

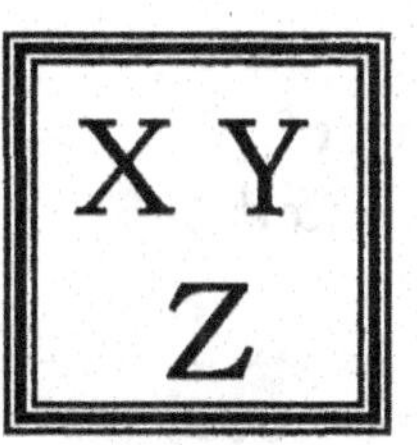

UNIT 5
Solving Equations

Section 5-1: Introduction

Previously, we used the function $A(t) = 275t + 17{,}218$ to model the amount of college debt as a function of the years since 2000 that the student graduated. What if we wanted to know when the amount of debt would be $20,000?

Here, we are asking what input t would produce an output *A(t)* equal to 20,000. Since $A(t) = 275t + 17{,}218$, this results in the equation

$$20{,}000 = 275t + 17{,}218.$$

In previous courses you have solved equations like this, but in this unit we will be solving equations in the context of functions. In addition, we will use the graphing capability and the Goal Seek and Solver tools of Excel to solve equations.

Section 5-2: Solving Equations Algebraically

Let's go back and review a little about solving equations. Two equations are **equivalent** if they have exactly the same solutions. For example, 2x + 1 = 5 is equivalent to 2x = 4 because they both have 2 as a solution. There is a series of operations that will allow you to change an equation into an equivalent equation.

<hr>

Operations for Changing Equations into Equivalent Equations

- **Add or subtract the same number or algebraic expression to both sides of the equation.**

- **Multiply or divide both sides by the same number.**
  Note: *You cannot divide both sides by 0.*

- **Combine like terms on either side of the equation.**

- **Interchange the two sides of the equation.**

<hr>

Example 5-2.1:   Solve the equation $20{,}000 = 275t + 17{,}218$.

Solution:

$$20{,}000 = 275t + 17{,}218$$

$$275t + 17{,}218 = 20{,}000 \quad \text{(Interchange the two sides of the equation)}$$

$$275t + 17{,}218 - 17{,}218 = 20{,}000 - 17{,}218 \quad \text{(Subtract the same number from both sides)}$$

$$275t = 2782$$

$$\frac{275t}{275} = \frac{2782}{275} \quad \text{(Divide both sides by 275)}$$

$$t \approx 10.1 \qquad \blacklozenge$$

The example above illustrates that when you solve an equation you are repeatedly finding equivalent equations until you get one in which the solution is obvious like, $t \approx 10.1$.

Example 5-2.2:   Given a function such as $f(x) = 4x + 8$, we previously were asked to find $f(4)$, but now we want to solve $f(x) = 4$. In other words, find the input when the output of $f(x)$ is 4.

Solution: We have

$$f(x) = 4x + 8$$

$$4 = 4x + 8 \quad \text{(Substitute 4 for } f(x))$$

$$-4 = 4x \quad \text{(Subtract 8 from both sides)}$$

$$x = -1 \quad \text{(Divide both sides by 4)}$$

So an input of -1 gives an output of 4. Let's check the solution:

$$f(-1) = 4(-1) + 8 = -4 + 8 = 4. \qquad \blacklozenge$$

Example 5-2.3:   Solve $7-(2x-3)=3x$

Solution: In this example we have the unknown, x, on both sides of the equation. Most people would first simplify the left side of the equation and then get all the unknowns on the same side of the equation. So let's do that.

$$7-(2x-3)=3x$$
$$7-2x+3=3x$$
$$10=3x+2x$$
$$10=5x$$
$$x=2$$

◆

Activity 5.1

Solve the following equations for x.

a) $5-(x+3)=1$

b) $2(x+4)=4x$

c) $3x-7=4-3(x+4).$

Example 5-2.4: Suppose you are planning a class trip to New York and the total cost for the trip is $3,000 for any number of students traveling (the limit is 200). In this situation the cost per student is a function of the number of students n. This can be expressed in function notation as

$$C(n) = \frac{3,000}{n}.$$

Note that the domain of the function is the set of integers from 1 to 200. How many students would have to attend so the cost per student is $70?

Solution: Here we are being asked, what do we input for n to get an output of $70? In other words, we are asked to solve the equation

$$\frac{3,000}{n} = 70.$$

Before we solve this equation, recall that if $\frac{A}{B} = C$ then $A = C \times B$.

Using this fact we have

$$\frac{3,000}{n} = 70$$

$$3,000 = 70n$$

$$n = \frac{3,000}{70} = 42.86 \approx 43$$

Therefore, 43 students would have to attend the trip so the cost per student is $70. ◆

Activity 5.2

Suppose that 6 students have already committed to going on the trip and that n represents the number of additional students to be recruited for the trip. Then the cost per student depends on n and can be represented by the function

$$C(n) = \frac{3,000}{6+n}.$$

Find the number of additional students that are needed so the cost per student is $60.

Recall the distance formula: *distance* is equal to *rate* times *time*, or

$$d = r \times t.$$

Suppose we wanted to solve this equation for r. We could do this by dividing both sides of the equation by t.

$$d = r \times t$$

$$\frac{d}{t} = \frac{r \times t}{t} \qquad \text{(Divide both sides by } t)$$

$$r = \frac{d}{t}$$

Similarly we could solve for t and get $t = \dfrac{d}{r}$.

The distance from Keene to Springfield, Massachusetts is 75 miles. So we can write trip time driving to Springfield from Keene as a function of average speed,

$$t(r) = \frac{75}{r}.$$

Activity 5.3

For the following questions:

    i) Use function notation to write an algebraic calculation or equation that can be used to answer the given question.

    ii) Solve the problem, and then write a sentence stating your answer to the original question.

a)   A bus makes the trip from Keene to Springfield at an average speed of 40 miles per hour. How long does it take to make the 75 mile trip?

b)   Driving a car you make the trip from Keene to Springfield in 1.5 hours. What was your average speed?

Example 5-2.5: Keene State College uses a computer system that processes data for an average day in 8 hours. In order to process data more rapidly the college plans to add new components to the system. One set of new components can handle the data in 5 hours, without the present system. How long would it take the new system, a combination of the present system and the new components to process the data?

Solution: Let x = the number of hours for the new system to process the data. The present system takes 8 hours to process the data, which means it would process $\frac{1}{8}$ of the data in one hour, or

$\frac{1}{8}x$ of the data in x hours. In the same way, the new components can process $\frac{1}{5}x$ of the data in x hours. When x hours have passed, the new system will have processed all of the data. Therefore,

$$\underbrace{\frac{x}{8}}_{\substack{\text{part of data processed}\\\text{by present system}}} + \underbrace{\frac{x}{5}}_{\substack{\text{part of data processed}\\\text{by new components}}} = \underbrace{1}_{\substack{\text{one completed process}}}$$

$$\frac{5x+8x}{40} = 1 \qquad \text{(Add fractions as shown in Unit 2)}$$

$$5x+8x = 40 \qquad \text{(Multiply both sides by 40)}$$

$$13x = 40 \qquad \text{(Combine like terms)}$$

$$x = \frac{40}{13} \approx 3.1 \qquad \text{(Divide both sides by 13)}$$

So the new system should take about 3.1 hours to process the data. ◆

Example 5-2.6: Solve $x + 3y - 9 = 0$ for y.

Solution:

$$x + 3y - 9 = 0$$

$$x + 3y - 9 + 9 = 0 + 9 \qquad \text{(Add 9 to both sides)}$$

$$x + 3y = 9$$

$$x - x + 3y = 9 - x \qquad \text{(Subtract x from both sides)}$$

$$3y = 9 - x$$

$$y = \frac{9 - x}{3} \qquad \text{(Divide both sides by 3)} \quad ◆$$

Example 5-2.7: Solve P = 4L + 2W, for W.

Solution: The goal is to get W = "something". To accomplish this you need to eliminate the term 4L and the factor 2 from the right side of the equation. Suppose you decided to eliminate the factor 2 first. Since 2 is a factor you have to divide **every** term on both sides of the equation by 2. That will give you

$$\frac{P}{2} = \frac{4L}{2} + \frac{2W}{2}$$

which simplifies to

$$\frac{P}{2} = 2L + W$$

Now you have to eliminate the term 2L from the right-hand side, so you subtract it from both sides since 2L is a term. Hence,

$$\frac{P}{2} - 2L = W .$$

Suppose your friend decided to solve for W by first subtracting 4L from both sides, getting

P - 4L = 2W.

Next, your friend divided both sides by 2 to get

$$\frac{P-4L}{2} = W .$$

Although it looks like you both got different answers, the two equations that you wrote are equivalent. If you separate your friend's expression for W into two fractions it will be the same as your expression or if you combine your expression for W into one fraction, you will see that it is the same as your friend's expression.

♦

Homework Section 5-2

1. Solve the following equations for the indicated variable.

    a) $3x + 9 = 30$ for x

    b) $4(2x + 7) - 5 = 40$ for x

    c) $\dfrac{26}{n - 5} = 12$ for n

    d) $5(d - 1) - 2d = 8$ for d

2. We saw in an earlier unit that the percent change is given by the formula

    $$100\left(\frac{New - Old}{Old}\right) = percent\ change.$$

    If we just wanted the fractional change we would use the formula

    $$\left(\frac{New - Old}{Old}\right) = fractional\ change.$$

    Suppose the population of a town was 21,000 in 2007 and the population increased by 0.05 from 2007 to 2009, what was the population in 2009? (Here *fraction change* = 0.05 and *Old* = 21,000.)

3. Solve the following equations for the indicated variable.

    a) $2x + 5y - 9 = 0$ for y

    b) $3x - 4y + 7 = 0$ for y

    c) $I = \dfrac{E}{R}$ for R

    d) $v = v_0 + at$, velocity of a falling object, for t

Section 5-3: Approximating Solutions to Equations Graphically

Some equations are difficult or cannot be solved algebraically. One way to at least get an approximation of the solution or find the number of solutions is to use a graph. Let's start with an example of an equation we can solve algebraically to illustrate the method.

Example 5-3.1:   Solve the equation $4x - 7 = 5$ graphically.

Solution: Think of both sides of the equation as functions. In other words, $f(x) = 4x - 7$ and $g(x) = 5$. We are looking for the input that makes $f(x) = g(x)$. Start by graphing both functions on the same axis and find the value of x where the two graphs intersect.

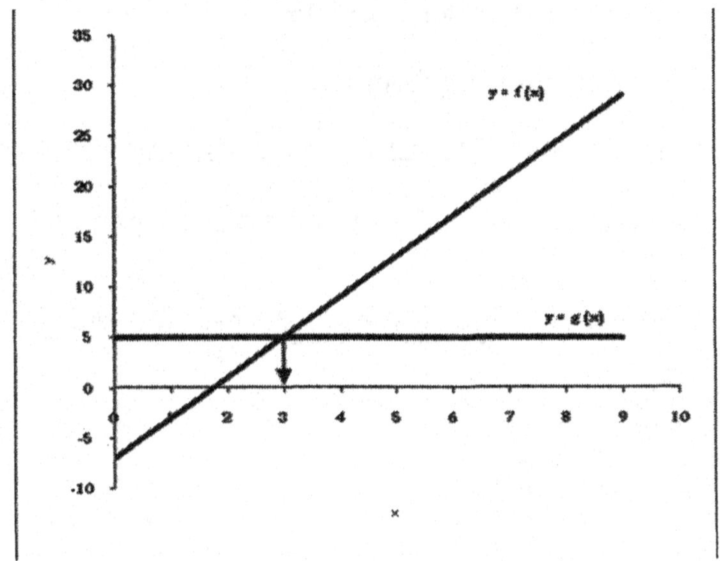

Looking at the graph you can see that the solution to the equation $4x - 7 = 5$ is $x = 3$.   ♦

Example 5-3.2:   Another method we can use is to get all terms on one side of the equal sign to form a function f(x) and then find solutions to $f(x) = 0$. Graphically the solution(s) will be where the graph of f(x) crosses the x-axis. Let's go back to the last example and solve $4x - 7 = 5$ using this method.

Solution: $4x - 7 = 5$

$4x - 7 - 5 = 5 - 5$ (Subtract 5 from both sides)

$4x - 12 = 0$

Now that all the terms are on one side of the equation, we can let $f(x) = 4x - 12$.

Using Excel to graph f(*x*), we will start by plotting the domain 1 to 20. The chart will be flexible enough for us to change this to other values if we need to. Perform the following steps on Excel.

In A2 type xStart and in A3 type xInc (signifies the increment by which *x* will increase).

In B2 type 1 and in B3 type 1.

In A5 type x and in B5 type f(x).

In A6 enter =B2, in A7 enter =A6+$B$3. Copy this down to row 25.

In B6 enter =4*A6-12. Copy this down to row 25.

In D1 enter the following formula.

="f(x) from x = "&MIN(A6:A25)&" to "&MAX(A6:A25)&""

Using the data for x and f(x), create a scatter plot as shown below.

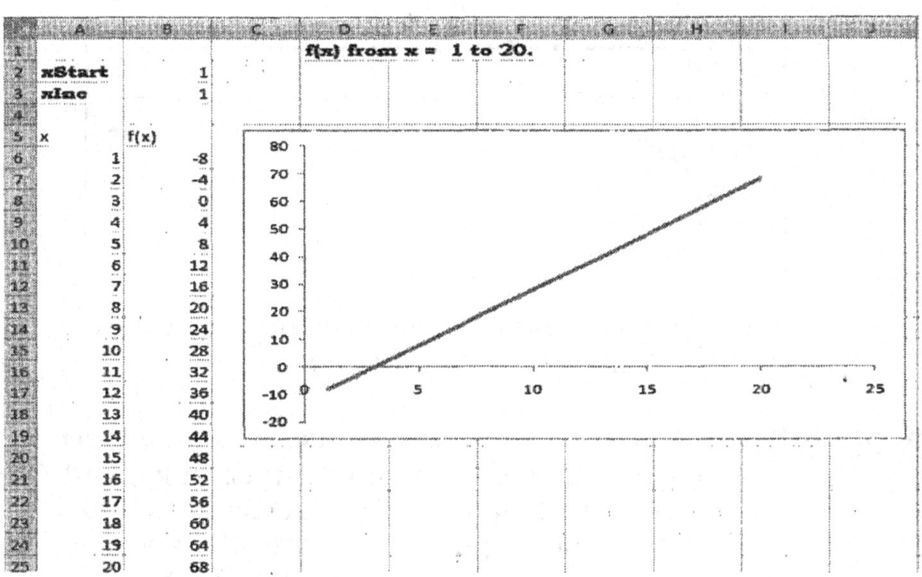

Recall that the solution to the equation is where the graph of f(x) crosses the x-axis. From the graph we see the solution is between 1 and 5. To get a better approximation let's change the domain to be from 1 to 5 by changing the value in B3 from 1 to 0.2.

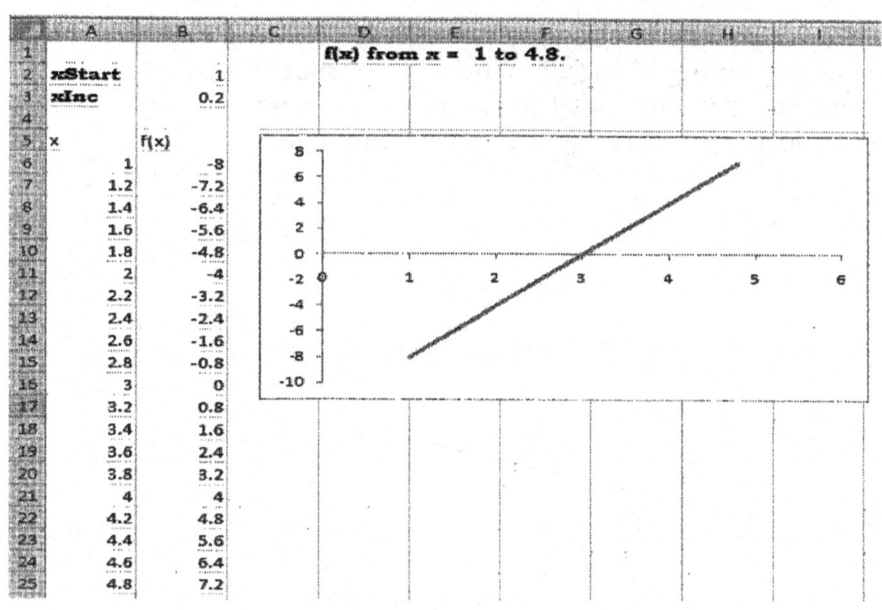

From this graph we can see that the solution is x = 3.

Example 5-3.3:   Solve the equation $2x^2 - 5 = 3.5x + 27$.

Solution: Letting $f(x) = 2x^2 - 5$ and $g(x) = 3.5x + 27$ we are looking for where the graphs of f(x) and g(x) intersect. Looking at the figure below, there are two points where the graphs intersect. One is at approximately x = 5 and the other is at approximately x = -3.

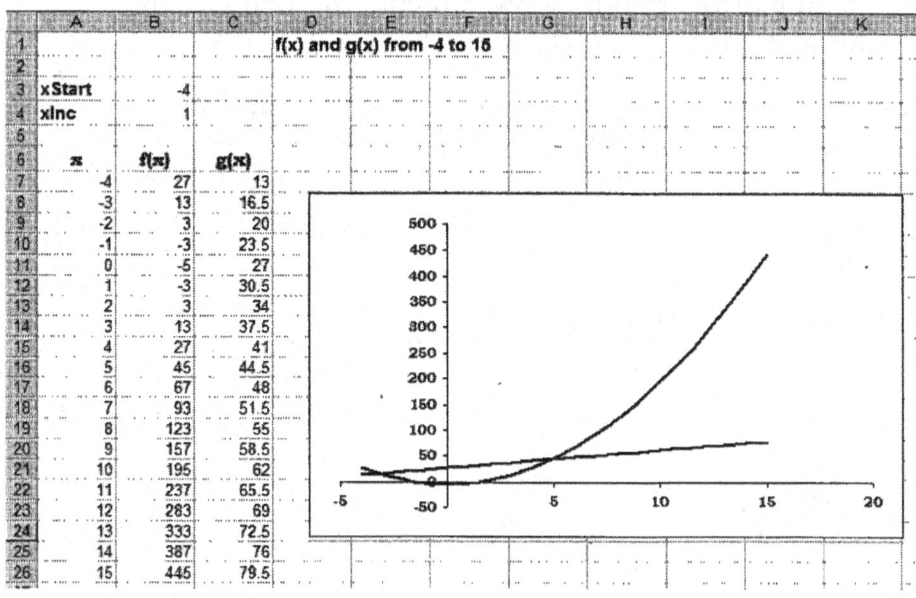

To get better approximations let's change the viewing domain. Notice xInc was changed to 0.2. When the cursor is placed at the point of intersection, the point (-3.2, 15.8) is shown. So –3.2 is an approximation to one of the solutions on the equation.

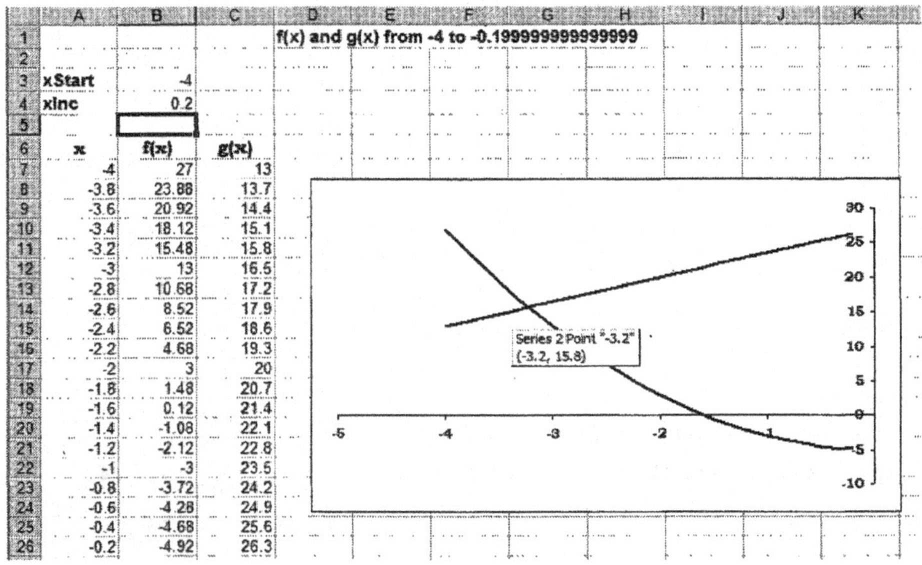

The other solution is approximately x = 5.   ◆

We could have also solved the equation $2x^2 - 5 = 3.5x + 27$ by first rewriting as $2x^2 - 3.5x - 32 = 0$. Graphing this function, we are looking for the values of x that give an output of 0. Looking at the graph below you can approximate the solutions, $x \approx -3.2$ and $x \approx 5$.

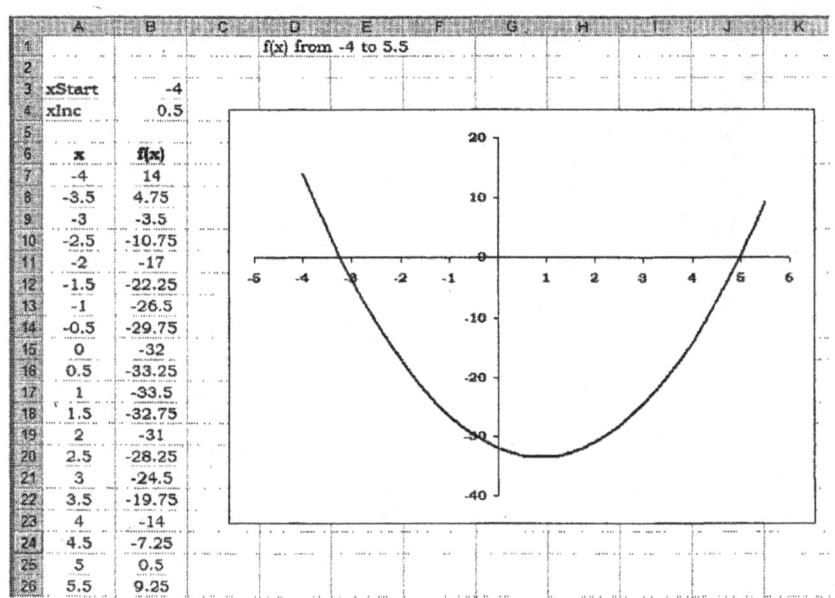

Activity 5.4

Given the graph of $y = f(x)$ below, approximate the solutions to the equation $f(x) = 0$ between 1 and 10.

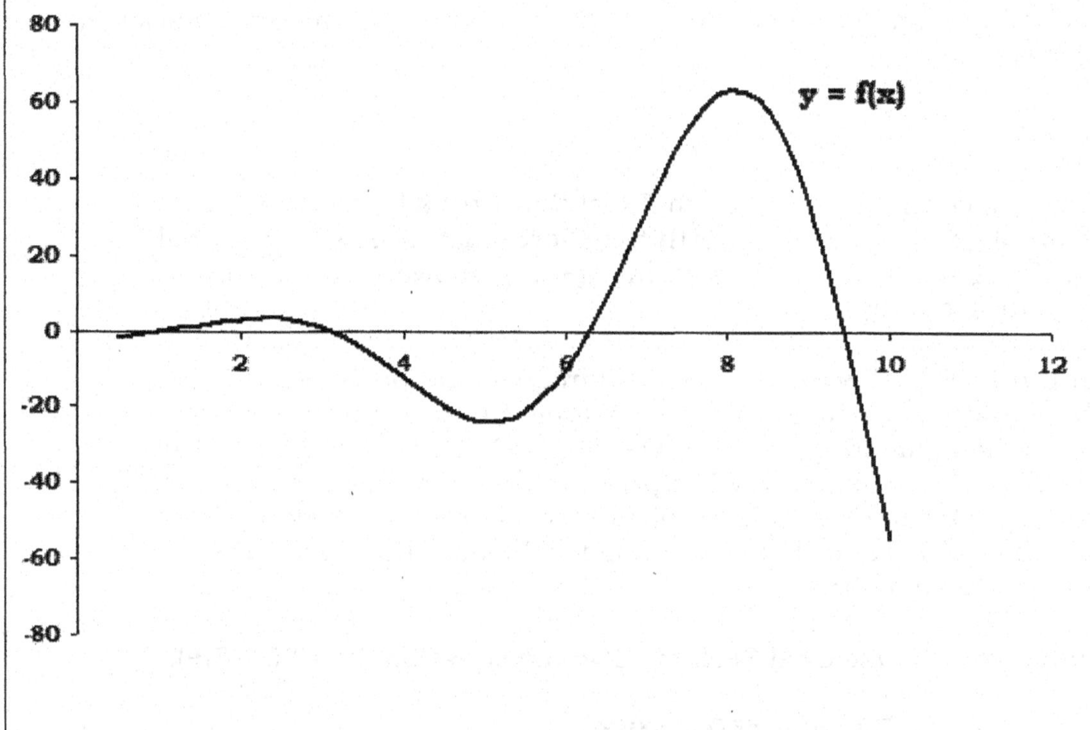

<u>**Section 5-4:   Using Excel's Goal Seek and Solver**</u>

In this section we will look at two ways that Excel can approximate the solution to an equation. Excel's Goal Seek and Solver operate by guessing a trial solution, trying it, and using the result to intelligently select a new trial solution. It is similar to the way you might use trial and error to find a solution. When we study systems of equations in a later unit we will also use Solver to solve these equations.

Goal Seek

Goal Seek is a computer program in Excel that works by trial and error to solve problems for you. It is worth the effort to learn how to use Goal Seek. By doing so you'll save lots of time getting answers, and also improve your accuracy.

You can think of Goal Seek as a special kind of equation solver which can solve equations of the form $f(x) = constant$. Suppose you know the equation of the demand curve to be Q = 50 - 5(P). Here Q is the quantity demanded, and P is the price. What price would make the demand equal to 40? Goal Seek can solve this problem by trial and error, taking different values for P and then searching for the one that makes the quantity very close to 40.

Here's how you can use Goal Seek to solve the equation 40 = 50 – 5(P).

1. In cell B2 enter **=50 – 5*B3**.
2. In cell B3 enter **3**. This is just an initial guess to the solution of the equation. You could enter another number if you wanted.
3. To find Goal Seek in Excel 2007, go to **Data -> What If Analysis -> Goal Seek**.

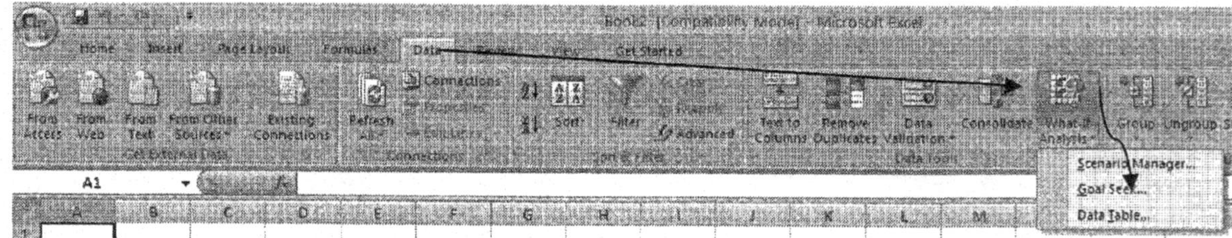

4. Enter B2 in the "Set cell:" box as shown on the next page.

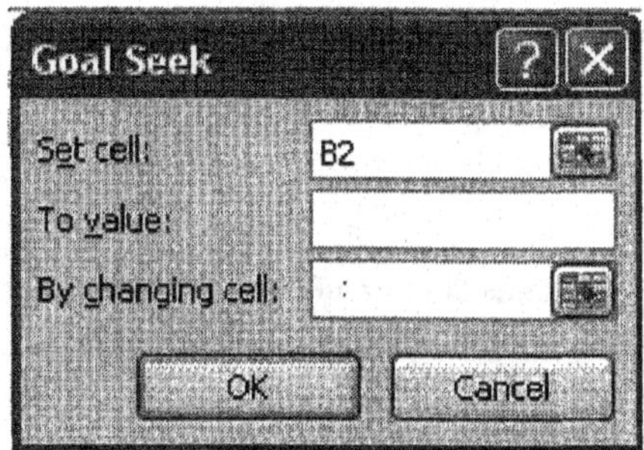

5. Using the mouse pointer, click on the "To value:" box, and enter 40. Don't click OK at this point.
6. Using the mouse pointer, click on the "By changing cell:" box and enter B3.
7. Click on OK.

Excel places the solution of 2 in cell B3.

Example 5-4.1:   Use Goal Seek to solve $2x - (3x + 7) = 10$.

Solution: Enter the following:

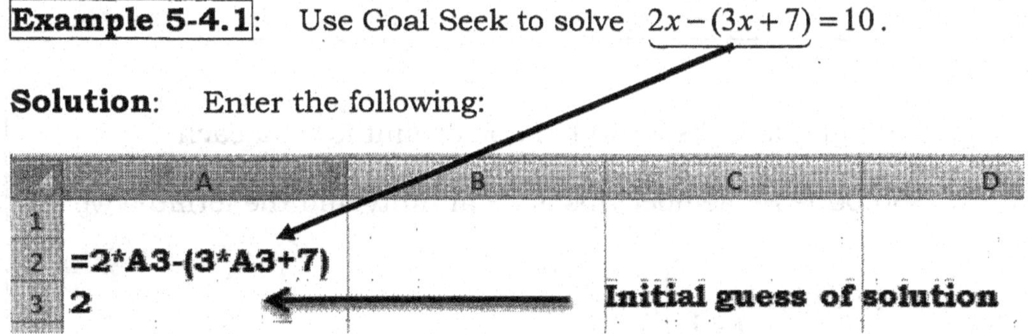

Go to **Data -> What If Analysis -> Goal Seek**.

Then enter the following in the Goal Seek box:

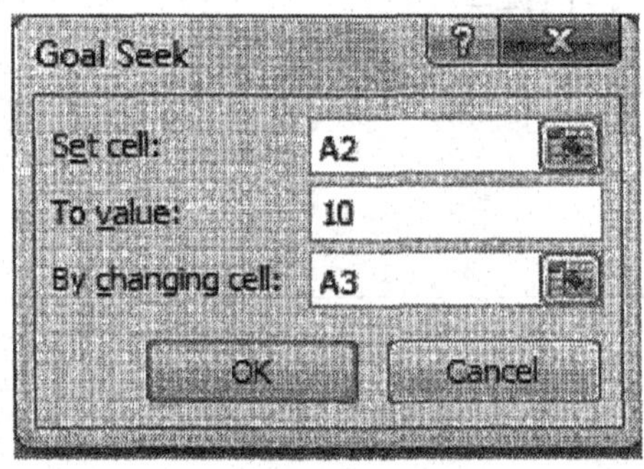

The answer will then appear in cell A3, $x = -17$.   ♦

As you can see from the last example, the Goal Seek dialog box displays three input boxes summarized below.

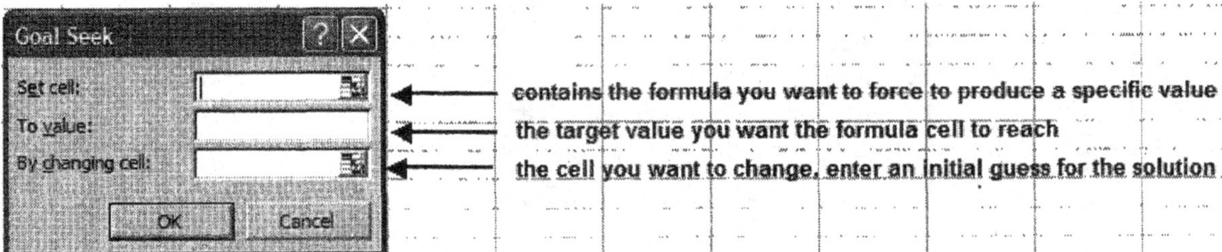

Let's look at a problem we previously looked at and solve the problem using Goal Seek. In a later unit we will learn to solve the problem without using Excel.

Activity 5.5

Suppose your grandfather deposits $3000 in an account for you each December 31st and the money earns 4% per year. How many years until the account is worth $50,000? Remember this is an annuity and the formula we used was

$$A(n) = a\left(\frac{r^n - 1}{r - 1}\right).$$

In this example $r = 1.04$ and $a = \$3,000$. We want to use Goal Seek to solve the following equation.

$$3000\left(\frac{1.04^n - 1}{1.04 - 1}\right) = 50,000.$$

Solver

Solver is a more powerful version of Goal Seek. Whereas Goal Seek has a single
target cell and changing cell, Solver can work out solutions to problems
involving several changing cells. In it, you can also set up constraint criteria. As
well as being able to find a given solution to an equation, Solver can find the
maximum or minimum value of a function. This is not only useful in
mathematics but is widely used in business to find maximum profits or
minimum costs.

Your computer may not have Solver already installed in Excel, so you may need
to add it before doing these examples. Use the help feature in Excel to find
directions for adding Solver.

Example 5-4.2:   Use Solver to solve the equation $3x + 5 = 23$.

Solution: This equation can easily be solved without using Solver or Goal
Seek, but we will use Solver just to get an idea how it works.

Cell A1 will be our changing cell; this is where our solution will
appear. In cell A2 we will enter =3*A1+5. Excel will change cell A1
until cell A2 is equal to 23.

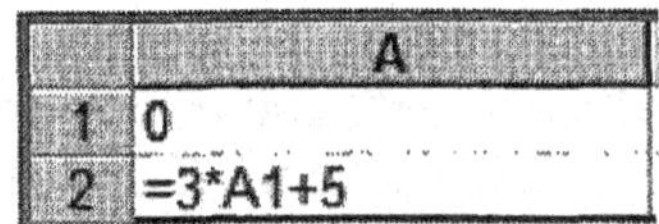

Now in the Data tab go to Solver. The following window will
appear:

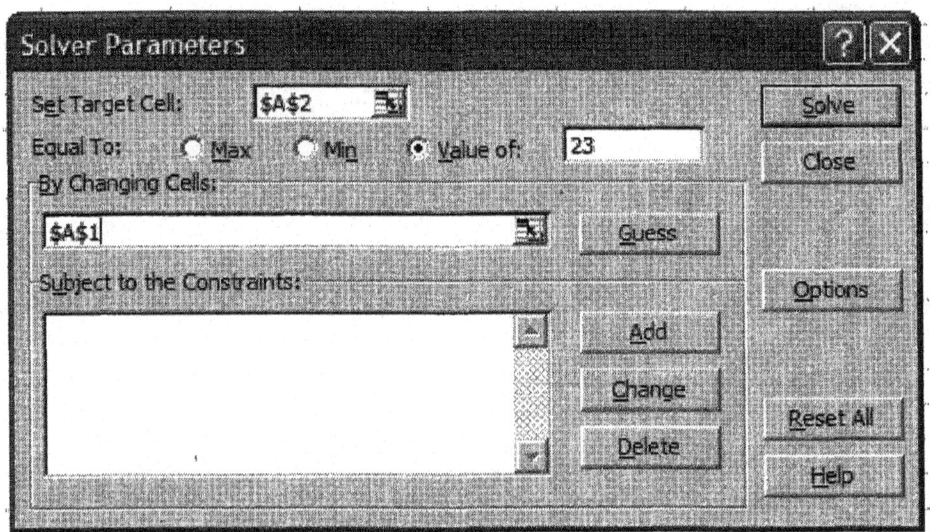

Enter $A$2 in Set target Cell, hit the Value of: button and enter
23, and in the By Changing Cells box enter $A$1. Now hit Solve,
and the solution should appear in cell A1.

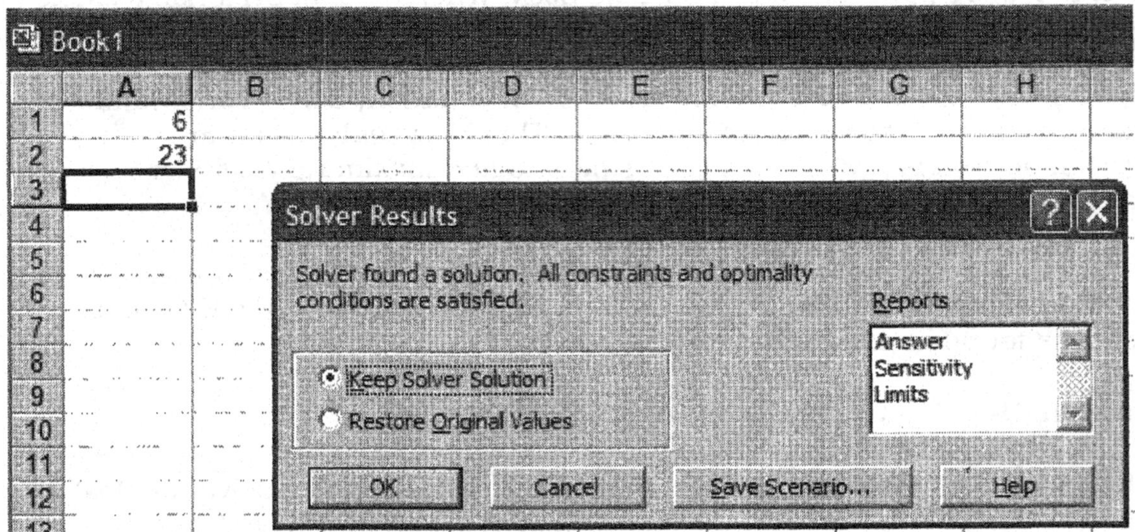

Notice a window appears that tells you that Solver found a solution. In this case the solution to the equation is x = 6.

♦

On the next page, let's look at an equation that we can't solve the way we have been solving equations and also has more than one solution. This example will also illustrate the value of looking at the graph of a function to help locate solutions to an equation.

Example 5-4.3:   Solve $x^3 + 3x^2 + 3 = 6$.

Solution: Think of this as asking what do you have to input into the function $f(x) = x^3 + 3x^2 + 3$ to get an output of 6. Let's look at the graph of the function to get an idea of what the solution(s) is.

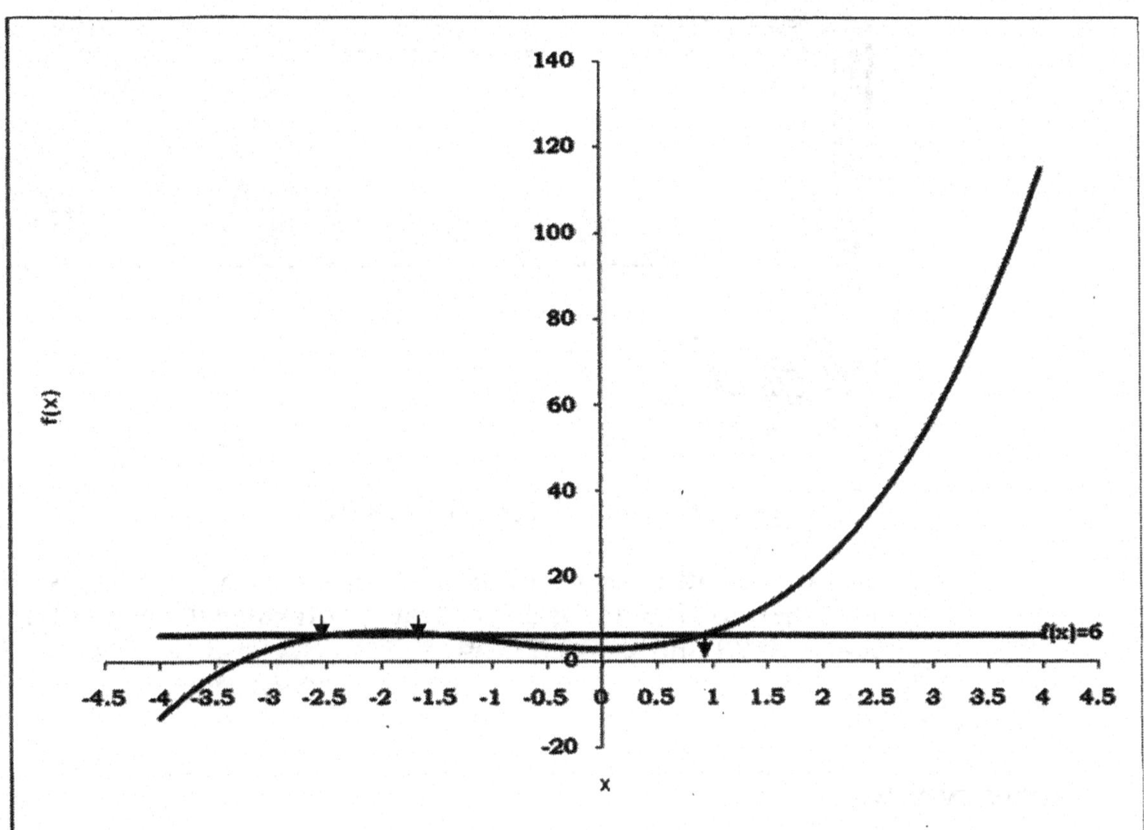

You can see from the graph of the function that there are three solutions to the equation. The three roots are close to 1, -1.5, and –2.5. We will use Solver three times using these values as our first guess to the solution.

First let's find the solution that is close to 1.

A	B	C
1	First guess to the solution	
=A1^3+3*A1^2+3		

Now use Solver.

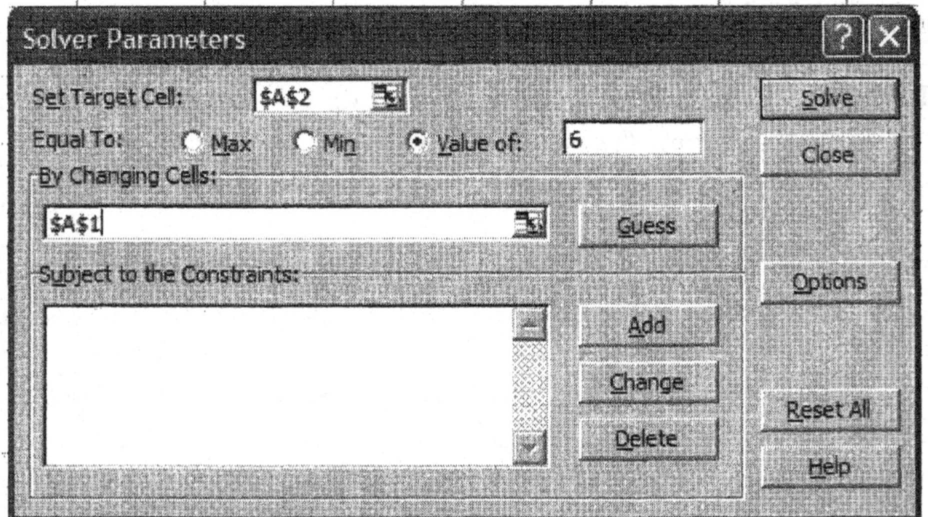

Solver gives the solution $x = 0.879385$.

To find the other two solutions, change the value in cell A1 to -1.5 and then to -2.5 using Solver twice. You will get the solutions $x = -1.3473$ and $x = -2.53209$.   ♦

Activity 5.6

Solve $2x^3 - 3x + 1 = 5$ using Solver.

Activity 5.7

In Unit 3 we looked at the following formula:

$$FV = PV(1+r)^n, \text{ where}$$

FV = future value
PV = present value
r = interest rate
n = number of years

So if you invested $1,000 at an annual interest rate of 3.5% the amount (future value) you would have after n years would be

$$A(n) = 1000(1.035)^n.$$

Suppose you want to know in how many years will the $1,000 grow to $5,000. Looking at the table and the graph below, we see that it will take about 47 years for you to have $5,000 in your account.

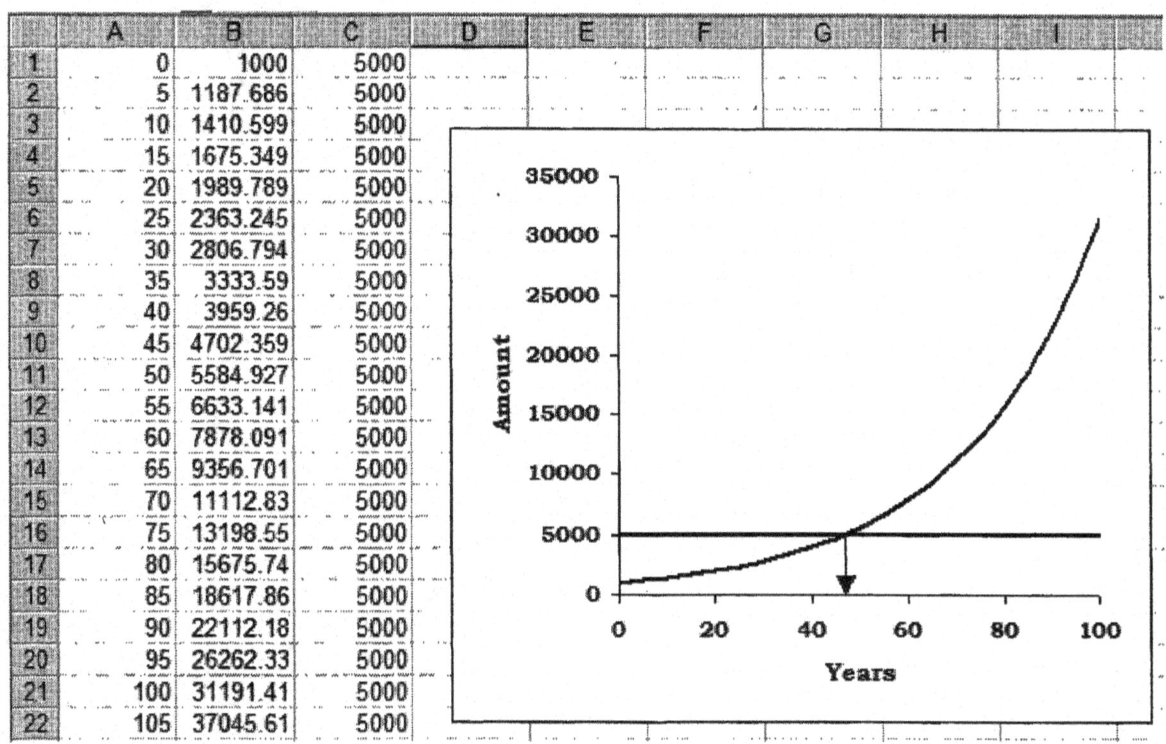

	A	B	C
1	0	1000	5000
2	5	1187.686	5000
3	10	1410.599	5000
4	15	1675.349	5000
5	20	1989.789	5000
6	25	2363.245	5000
7	30	2806.794	5000
8	35	3333.59	5000
9	40	3959.26	5000
10	45	4702.359	5000
11	50	5584.927	5000
12	55	6633.141	5000
13	60	7878.091	5000
14	65	9356.701	5000
15	70	11112.83	5000
16	75	13198.55	5000
17	80	15675.74	5000
18	85	18617.86	5000
19	90	22112.18	5000
20	95	26262.33	5000
21	100	31191.41	5000
22	105	37045.61	5000

Use Solver to solve the problem.

Unit 5 Review Exercises

1. Solve the following equations. When necessary, round the answer to the nearest tenth.

    a) $4x - 7 = 41$ for x

    b) $6 - 4(t + 2) = -26$ for t

    c) $\dfrac{200}{n-5} = 3$ for n

    d) $\dfrac{65}{r} = 1.5$ for r

    e) $\dfrac{x}{5} + \dfrac{2x}{3} = 2$ for x

    f) $3x - y + 10 = 0$ for y

2. Use Excel to graphically solve the following equations for x.

    a) $x^3 - 3x + 6 = 10$

    b) $x^3 + \sqrt{x} + 7 = 2$

3. Solve the equations in exercise 2 using Solver and Goal Seek on Excel.

4. Use Solver to solve the equation $1000(1.045)^n = 8500$.

5. Use Goal Seek to solve the equation $2500\left(\dfrac{1.02^n - 1}{1.02 - 1}\right) = 15,000$.

UNIT 6
Linear Functions

Section 6-1: Introduction to Linear Functions

Linear functions arise quite naturally in "real life." These are functions that represent a constant increase or a constant decrease. For example, one application of linear functions is called **straight-line depreciation**. Under straight-line depreciation, the value of an item, such as an automobile is assumed to drop by the same amount each year. Look at the following input-output table where P is a function of t.

t (input)	0	1	2	3	4
P (output)	1000	900	800	700	600

You can see from the table that **P** is decreasing at the constant rate of 100 for each increase of 1 in **t**. We say that 100 is the **rate of change** of P versus t. If you graphed this data you would see that the points all lie on a straight line and if you found additional points based on the pattern in the table, you would see that the new points also lie on the same line.

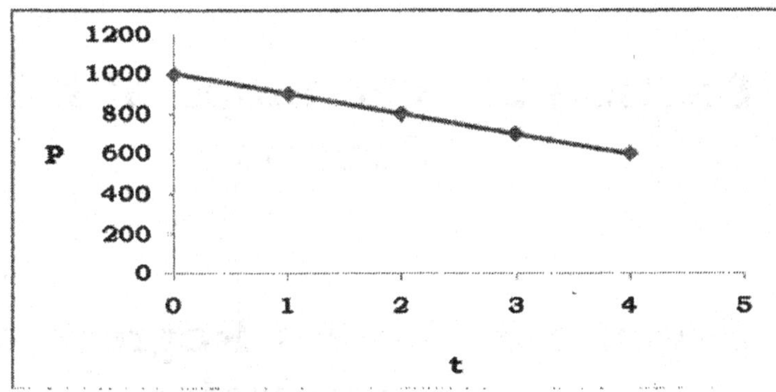

In general:

A **linear function** describes a quantity that changes at a constant rate. The graph of a linear function is a straight line. Numerically, a linear function is a function where the ratio, $\dfrac{change\ in\ output}{change\ in\ input}$, is constant over the entire domain of the function.

The symbol Δ is commonly used to denote change in a variable. We can therefore refer to the change in x as Δx (read "delta x"). Then,

$$\frac{\Delta y}{\Delta x} = \frac{change\ in\ output}{change\ in\ input}$$

denotes the rate of change for any linear function.

The federal income tax regulations for 2008 state that if you are married and filing jointly and your 2008 income was equal to or greater than $372,950, then your federal income tax is $100,894 plus 35 percent of the amount over $372,950. The tax for various incomes greater than $372,950 is given in the following table.

Income	372,950	373,950	374,950	375,950	376,950	377,950
Tax	100,894	101,244	101,594	101,944	102,294	102,644

Thinking of income as the **independent** (input) variable and tax as the **dependent** (output) variable, you can see that every increase of $1,000 in a person's income causes the tax to increase $350. It doesn't matter whether the income is $372,950 or $500,000. The **rate of change** of tax with respect to income is a constant number, $\dfrac{350}{1000}$. **It is the constant rate of change that makes a functional relationship linear**. Thus tax is a **linear function** of income.

Graphing the table given above, we get the following:

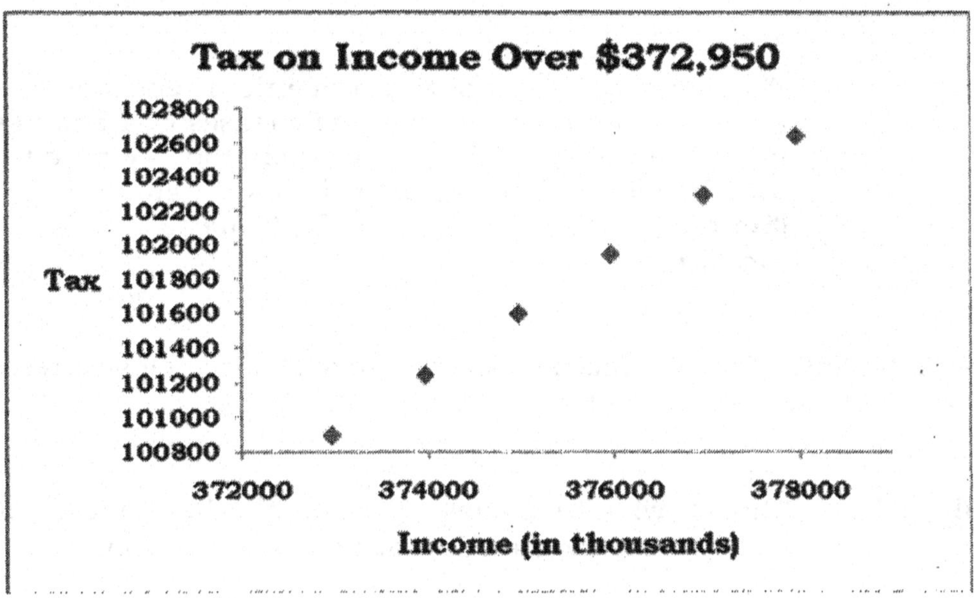

The graph shows that for every increase of $1000 in income, the tax increases by $350. We say that the **rate of change** of the tax with respect to income was $350 per $1000. We can write this formally as

$$\frac{\text{change in tax}}{\text{change in income}} = \frac{\Delta T}{\Delta I} = \frac{\$350}{\$1,000} = \$350 \text{ per } \$1000 \text{ increase in income.}$$

 According to the standardized growth and development charts used by many American pediatricians, the **median** weight for girls during their first six months of life increases at an almost constant rate. Starting at 7.0 lbs. at birth, for each additional month of life the female median weight increases by 1.5 lbs. Thus 1.5 is the **rate of change** in pounds per month. The table and graph below show two representations of the function. Note that the domain (input values) is between 0 and 6 months.

Age (in months)	Median weight for girls (in lbs)
0	7.0
1	8.5
2	10.0
3	11.5
4	13.0
5	14.5
6	16.0

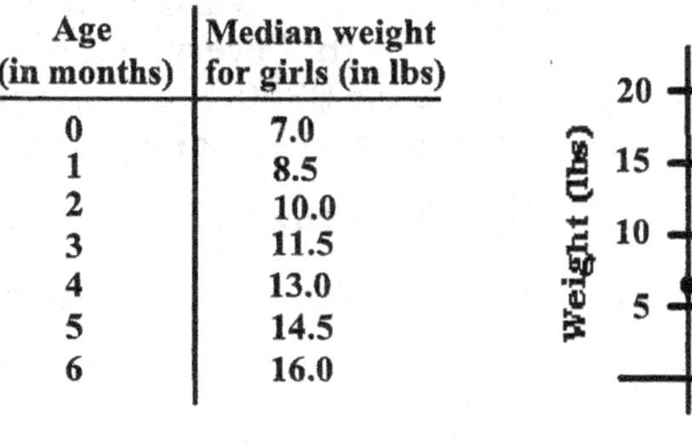

The rate of change, 1.5 lbs per month, means that as age increases by one month, weight increases by 1.5 lb. If we move 1 unit to the right on the graph, then we need to move up 1.5 units to find the next point on the graph.
(**Source:** L. Kime and J. Clark, *Explorations in College Algebra*) ◆

The idea that a linear function has a constant rate of change helps us recognize when a set of data describes a linear function.

Example 6-1.2: The following data show a boy's height, h, as a function of age, A. Does the data describe a linear function?

A (years)	h (inches)
2	33.0
3	35.8
4	38.6
5	41.4

Solution: If the data represents a linear function, the rate of change should be constant. The change in A is constant, 1, so if the function is linear the change in h, called the first difference, should be constant.

A (years)	h (inches)	first difference
2	33.0	
		35.8 - 33.0 = 2.8
3	35.8	
		38.6 - 35.8 = 2.8
4	38.6	
		41.4 - 38.6 = 2.8
5	41.4	

Since the first difference is constant, the rate of change is constant and the data represents a linear function.

Remember we can write this as

$$\frac{\text{change in h}}{\text{change in A}} = \frac{\Delta h}{\Delta A} = \frac{2.8 \text{ inches}}{1 \text{ year}}.$$

So the rate of change is 2.8 inches per 1 year. Every year a boy's height increases by 2.8 inches.

♦

Activity 6.1

Does the following data represent a linear function?

x	0	2	4	6
y	2	8	14	20

The **slope** of a linear function $y = f(x)$, denoted by m, can be calculated from values of the function at two points, x_1 and x_2, using the formula

$$Slope = \frac{Rise}{Run} = \frac{\Delta y}{\Delta x} = \frac{f(x_2) - f(x_1)}{x_2 - x_1}.$$

The slope is the constant rate of change that is characteristic of a line.

Example 6-2.1: Find the slope of the line joining the points (3, 5) and (9, 17).

Solution: According to the formula above,

$$\text{slope} = \frac{17-5}{9-3} = \frac{12}{6} = \frac{2}{1} = 2$$

The slope is $\dfrac{2}{1}$, which means for each unit increase in x, the y-value increases by 2. So if you are at any point on the line, if you increase the x coordinate 1 and the y coordinate 2, you will be at another point on the line.

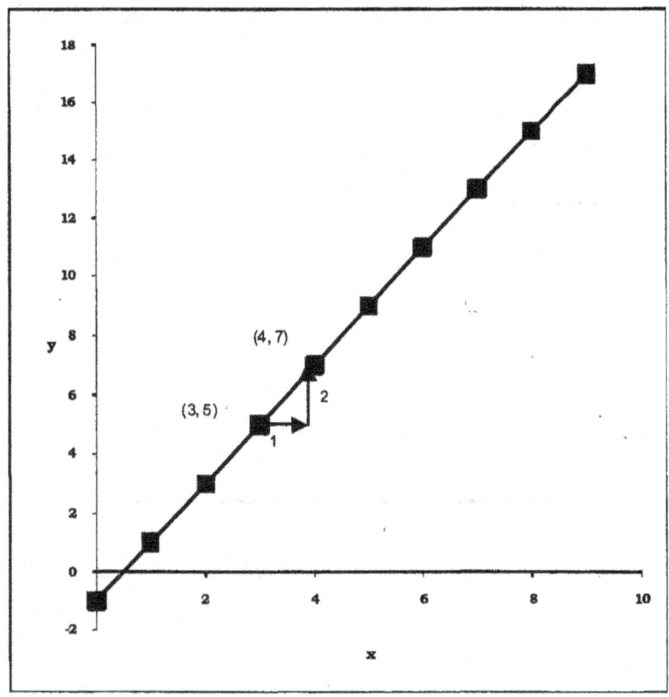

The slope of a line characterizes the "steepness "of the line. When the slope of a linear function is negative the graph will be falling as you go from left to right. The opposite will be true when the slope is positive, the graph will be rising as you go from left to right.

Example 6-2.2: In the figure below, lines B and C have positive slopes (the lines rise from left to right) and A has a negative slope (falls from left to right).

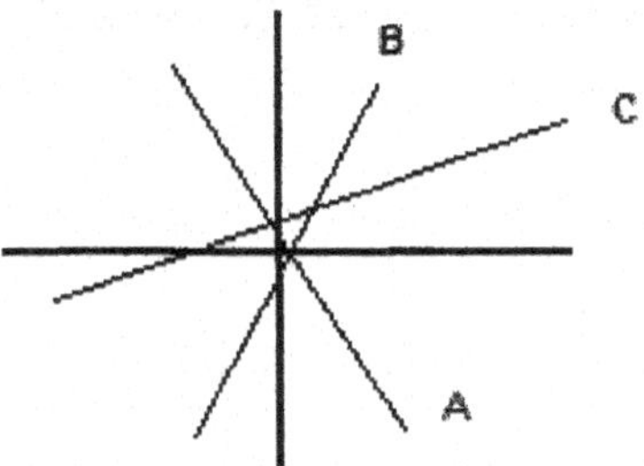

In addition, remember the slope measures the steepness of a line. If you take the absolute values of the slope, the larger the absolute value the steeper the line. Looking at lines B and C above, since B is a little steeper than C, the absolute value of its slope would be larger.   ♦

We also want to be able to write the linear function that passes through the two points. To do this we first have to know the general form of a linear function.

Linear Functions in General

A linear function has the form

$$y = f(x) = mx + b$$

and a graph where

m is the **slope**, or *rate of change of* y with respect to x
(0, b) is the **vertical intercept** or value of y when x is zero.

To find the linear function through two points, (x_1, y_1) and (x_2, y_2) we can use the following procedure:

1. Find the slope using $\dfrac{\Delta y}{\Delta x} = \dfrac{y_2 - y_1}{x_2 - x_1}$.
2. Substitute the slope and the x and y values from either known point into the equation $y = mx + b$.
3. Solve the resulting equation for b.
4. Write the linear function using the calculated m and b.

Example 6-2.3: Write the linear function that passes through the points (3, 5) and (9, 17).

Solution: We first find the slope. Using the slope formula, we have

$$m = \frac{y_2 - y_1}{x_2 - x_1} = \frac{17 - 5}{9 - 3} = 2.$$

Using the point (3, 5) and slope of 2, we substitute into $y = mx + b$ to get

$$5 = 2(3) + b$$

$$5 = 6 + b$$

$$-1 = b$$

So, m = 2 and b = -1. Substituting those values into $f(x) = mx + b$,

gives us the linear function $f(x) = 2x - 1$. ♦

Example 6-2.4: Suppose you can rent a van from Keene State College for an initial charge of $30 plus 40 cents per mile. Let C be the function whose input is n, the number of miles driven, and whose output is $C(n)$, the total cost.

1. Explain why this is a linear function.

2. Write the cost as a linear function of n.

3. Determine the cost of renting the van if you drove it 275 miles.

Solution:
1. This is a linear function since the rate of change (in this case, the charge per mile) is constant.

2. To find the function, we need the slope and the vertical intercept. The slope, or rate of change, is 40 cents or 0.4 dollars per mile. The vertical intercept is the cost if zero miles were driven which is $30. So the function is

    $$C(n) = 0.4n + 30$$

    where $C(n)$ is the total cost (in dollars) and n is the number of miles driven.

3.   To determine the cost, substitute 275 for n, and
      calculate:

$$C(275) = 0.4(275) + 30 = 140$$

So, the cost of renting the van would be \$140 if you
drove 275 miles.   ♦

Activity 6.2

Suppose a case of fresh strawberries cost \$10. Each day the strawberries sit in
the store 50 cents is taken off the price. What linear function would model the
cost of the strawberries? Using your linear function, how much would a case of
strawberries cost if they sat in the store for 7 days?

Activity 6.3

Write the slope of the function below as a ratio and give an interpretation of the
slope.

P = -2.5x + 150, where x = demand of a product and P. = price charged

---

Homework Sections 6-1 and 6-2

1. What two pieces of information determine a line?

2. Find the slope of the line joining the points (3, 7) and (11, 5).

3. Let $f(x) = 3x + 5$.

    a) What is the slope?

    b) What is the vertical intercept?

    c) What is $f(0)$?

    d) What is $f(1)$?

    e) What is $f(-2)$?

4. Find the equation of the line through the points (2, 5) and (9, 6).

5. Determine if each of the following tables represent a linear function. If so, find an equation for the line.

    a)

x	y
-2	-5
-1	-3
0	-1
1	1
2	3

    b)

x	y
-2	-5
-1	-2
1	4
3	10
6	19

6. Give a linear function that will convert yards to inches. (1 yard = 36 inches)

7. Suppose the linear function $y = 2{,}700x - 1{,}800$ represents the average salary y of an American citizen who has finished x years of school.

    a) Explain the meaning of the slope and vertical intercept in this function. (Use appropriate units in your explanation.)

    b) What does the linear function predict for the salary of a person who has completed 14 years of school?

Typical problems from algebra texts ask questions such as the following:

> A manager of a weekend flea market knows from past experience that if she charges x dollars for a rental space at the flea market, then the number y of spaces she can rent is given by the equation $y = 200 - 4x$. Sketch a graph of this linear equation. (*Algebra and Trigonometry* by Stewart, Redlin, and Watson, 2001, Brooks/Cole, page 114)

In this section we want to understand how to get the function $y = 200 - 4x$. This function is called a mathematical model, so we want to be able to build the model.

We will start with a table of values. Sometimes if the data fits a linear function exactly you may be able to write the linear function by observation. Unfortunately the data you collect will rarely fit a linear function exactly. In this case we will use a technique called "linear regression" to find the line that "best fits" the data. Excel will handle the calculations for us. Let's go through the process of linear regression by looking at an example.

Example 6-3.1: Build a model (linear function) to predict the number of wins by major league baseball teams.

Solution: We know what the dependent variable is, the number of wins, but we will have to hypothesize what the independent variable is. Let's try as the independent variable the difference between runs a team scores and the runs they allow. The following data is available on mlb.com:

	Team	Runs Scored	Runs Allowed	Wins
1	Team	Runs Scored	Runs Allowed	Wins
2	Yankees	915	753	103
3	LA Angels	883	761	97
4	Boston	872	736	95
5	Dodgers	780	611	95
6	Philadelphia	820	709	93
7	Colorado	804	715	92
8	St. Louis	730	640	91
9	San Francisco	657	611	88
10	Minnesota	817	765	87
11	Florida	772	766	87
12	Texas	784	740	87
13	Detroit	743	745	86
14	Atlanta	735	641	86
15	Seattle	640	692	85
16	Tampa Bay	803	754	84
17	Milwaukee	785	818	80
18	White Sox	724	732	79
19	Cubs	707	672	78
20	Cincinnati	673	723	78
21	Toronto	798	771	75
22	Oakland	759	761	75
23	San Diego	638	769	75
24	Houston	643	770	74
25	Mets	671	757	70
26	Arizona	720	782	70
27	Cleveland	773	865	65
28	Kansas City	686	842	65
29	Baltimore	741	876	64
30	Pittsburgh	636	768	62
31	Washington	710	874	59

We need to compute *runs scored - runs allowed*, which we will call difference. Highlight column D and then while in the Home tab go to Insert and then Insert Sheet Column.

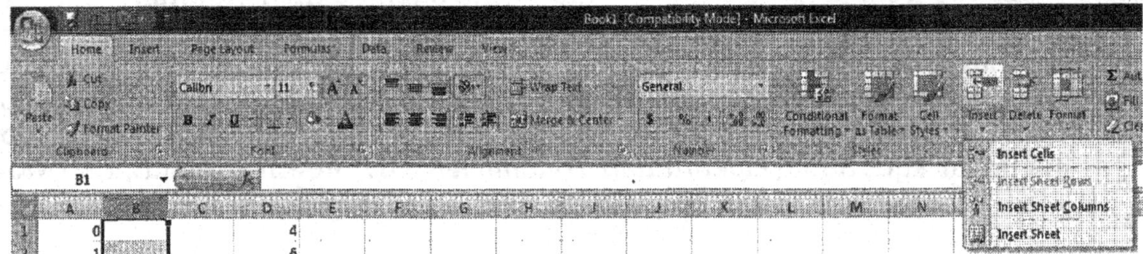

Enter the formula **=B2 - C2** in cell D2 and hit enter. Next click on cell D2 and then on the small square in the lower left corner and drag the formula down to cell D31. This will give you the run difference for all the teams.

We now want a scatter plot of difference versus wins. If there is a linear relationship between the variables, the data should fall in a linear pattern. Looking at the scatter plot it appears reasonable we could use a linear function to model the data.

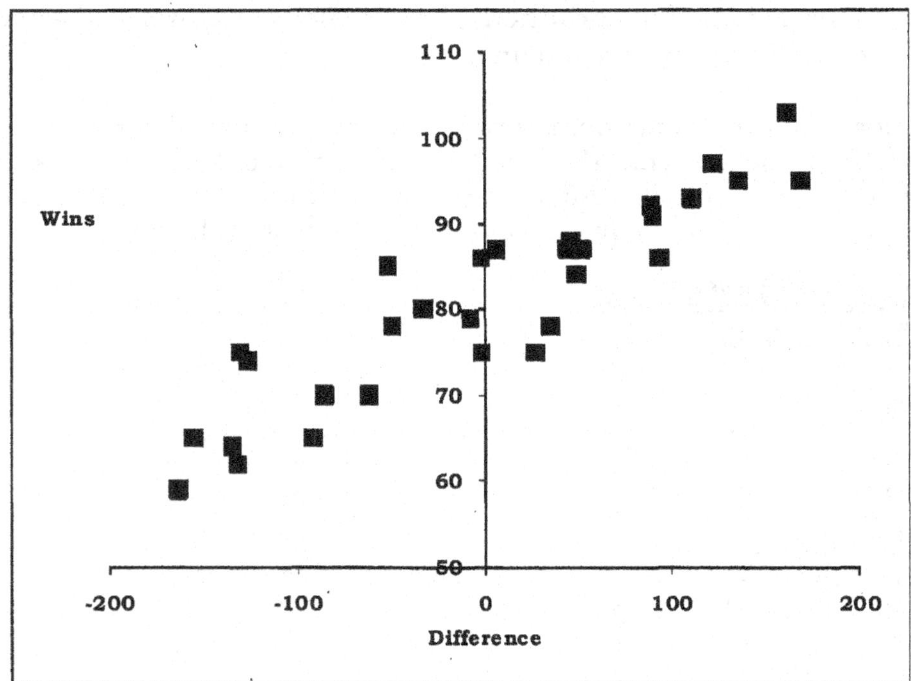

Now we want Excel to calculate the linear function. To do this right-click any of the data points on the graph, then select **Add Trendline**. The following will appear:

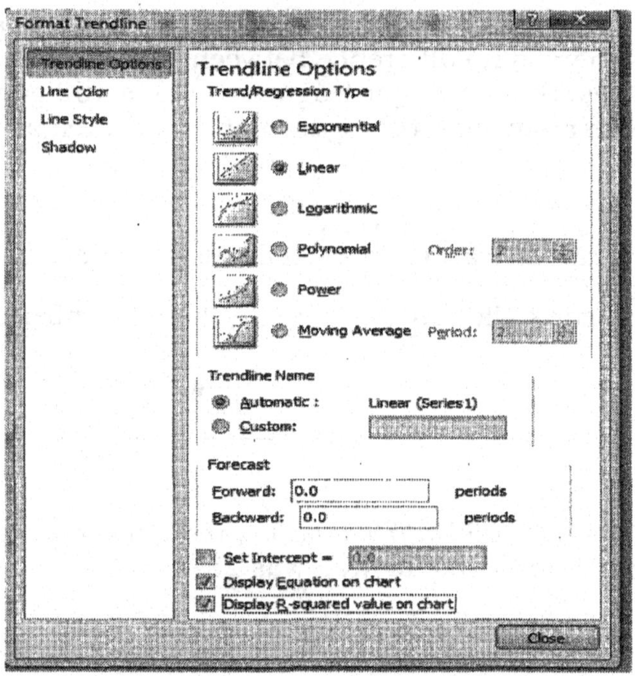

Select **Linear** and check **Display Equation on chart** and **Display R-squared value on chart** which appear at the bottom of the display. Hit Close.

Excel will add a line to the graph as well as the equation of the line. In addition it adds a value for r^2, **coefficient of determination**, which is a measure of how well the line fits the data. Here the value is 0.836 which is very good.

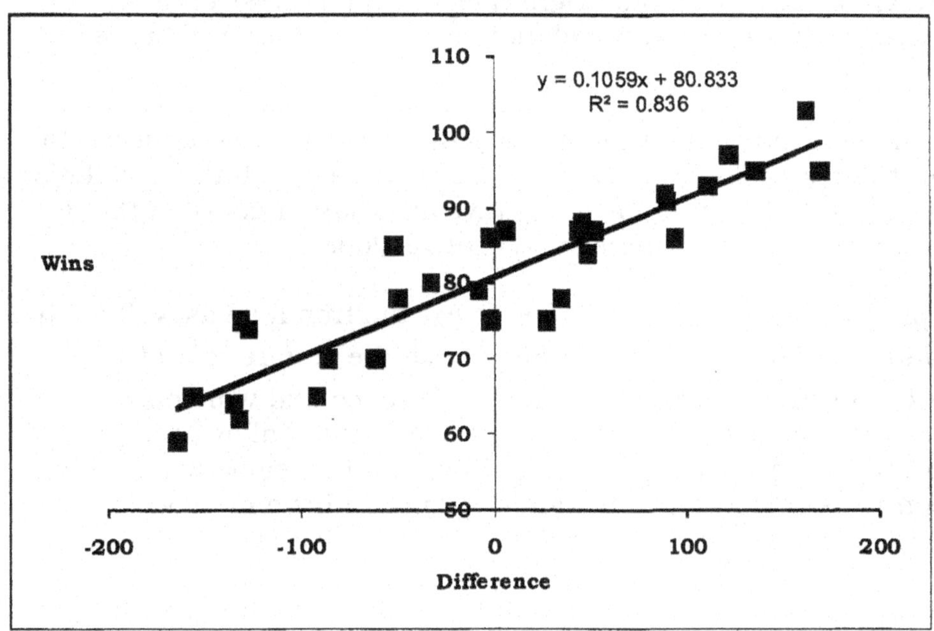

The equation we found is $y = 0.1059x + 80.833$ which in terms of the variables we are using is

$$wins = 0.1059*diff + 80.833.$$

Once we get this equation we need to determine if it seems reasonable. The vertical intercept is 81 and this tells us that if there is no difference between runs scored and runs allowed a team would win 81 games. Since a major league season is 162 games it seems reasonable that a team would win about half their games if they scored as many runs as they allowed.

The slope is 0.1059 and thinking of this as the ratio $\dfrac{0.1059}{1}$ tells us that every time the difference between runs scored and runs allowed increases by 1 the number of wins increases by 0.1059.

Activity 6.4

How many wins would the function *wins = 0.1059*diff + 80.833* predict for a team where the difference between runs scored and runs allowed was 25?

The standard measure of how well a regression line fits a set of data is called the **correlation coefficient**, or just correlation. This value measures both the direction and strength of a linear relationship between two numerical variables. The correlation is a number between -1 and 1 and is denoted by r. When $r = 1$ or $r = -1$, the data all lie exactly on a straight line and the regression line perfectly predicts the outputs. If the correlation is equal to 0, then there is little correlation.

If r is sufficiently larger than 0, then there is a positive association between the two variables and the regression line will have a positive slope. If the correlation is sufficiently less than 0, then there is a negative association between the two variables and the regression line will have a negative slope.

In the last example you saw how to use Excel to get the trendline as well as the linear function and r^2. To get the correlation, r, take the square root of r^2. In the example $r^2 = 0.836$ so the correlation of the difference and wins data is $r = 0.914$. This means that the difference (runs scored - runs allowed) and win data is highly correlated and has a positive association, i.e. teams where the difference between runs scored and runs allowed is large have more wins.

Below are two different scatter plots. The figure on the left shows the correlation of the number of home runs to the number of wins and the figure on the right shows the graph from the previous example. Even though the figure on the left has a much lower correlation, there is still a positive association, i.e. teams that hit more home runs have more wins.

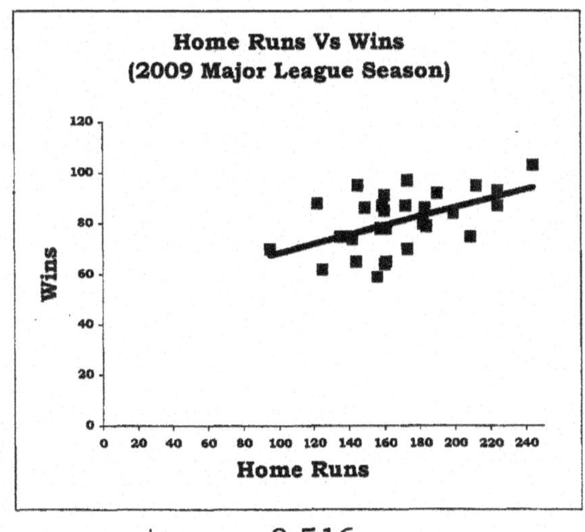

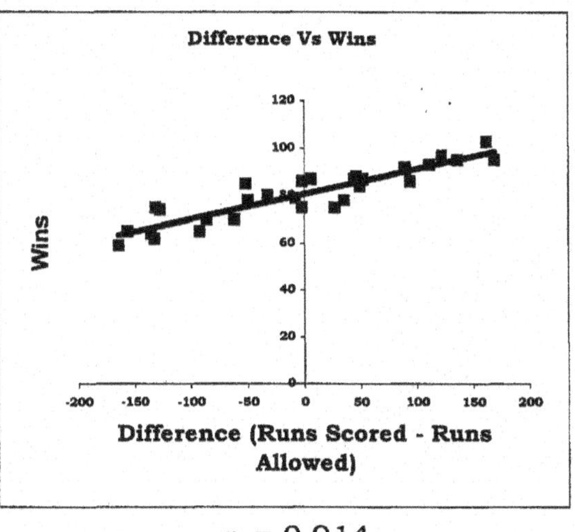

$$r = 0.516 \qquad\qquad r = 0.914$$

The **coefficient of determination**, r^2, measures the proportion of variation in the dependent variable that is explained by the independent variable in the regression model. In the baseball example $r^2 = 0.84$ which means that 84% of the variation in number of wins can be explained by the variability in the difference between runs scored and runs allowed. Only 16% of the variability in wins can be explained by factors other than what is accounted for by the linear model that uses only the difference in runs scored and runs allowed.

Correlation is often confused with causation. Two variables can be highly correlated without there being any causal relationship. For example, shoe size is strongly correlated to mathematics scores among elementary school children. This does not mean that big feet cause high math scores. The correlation is a result of the children growing and learning more as they age. Correlation is just a measure of the relationship between data. To establish causation, you would have to provide proof why change in one variable causes a corresponding change in the other variable.

Unit 6 Review Exercises

1. Without graphing determine if the following data represent a **linear** function? Explain your answer.

X	5.2	5.3	5.4	5.5	5.6
Y	27.8	29.2	30.6	32.0	33.4

2. Find the slope of the line through the given points.

   a) (2, 5) and (6, 18)   b) (-1, 5) and (3, -6)

3. Write an expression of the linear function which passes through the following points.

   a) (-2, 7) and (5, 28)   b) (3, 7) and (-1, 5)

4. Write the equation of the following function.

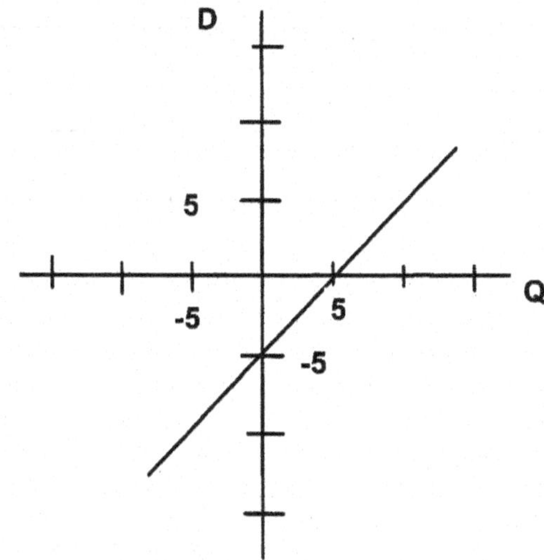

5. The data is from *Introduction to the Practice of Statistics* by Moore and McCabe. The children's heights were measured monthly over several years as part of a study of nutrition in developing countries. Use Excel to find the equation of a linear function that approximates this data. In addition find and interpret r^2, the coefficient of determination.

Age (months)	18	19	20	21	22	23	24	25	26	27	28	29
Height (cm)	76.1	77	78.1	78.2	78.8	79.7	79.9	81.1	81.2	81.8	82.8	83.5

6. The best way to understand the U.S. economy is by looking at **Gross Domestic Product (GDP)**, which is the statistic used to measure the economy. In other words, the U.S. economy, as measured by GDP, is everything produced by all the people and all the companies in the U.S.

GDP is important for three reasons, it is used to:

- determine if the U.S. economy is growing more quickly or more slowly than the quarter before or the same quarter the year before.

- compare the size of economies throughout the world.

- compare the relative growth rate of economies throughout the world.

The GDP affects you because if the GDP growth rate is speeding up, the Federal Government may raise interest rates to stem inflation. If GDP is slowing down then usually there are layoffs and unemployment.

Below is a table that gives the GDP (in billions) from 2000 to 2006. Use Excel to find the linear function that approximates this data. In addition, interpret the slope and find and interpret r^2. Use the linear function to predict the GDP for 2009. (The actual value was $14,905.05 billion.)

Years since 2000, t	0	1	2	3	4	5	6
GDP (billions)	9,951.5	10,286.2	10,642.3	11,142.1	11,867.8	12,638.4	13,398.9

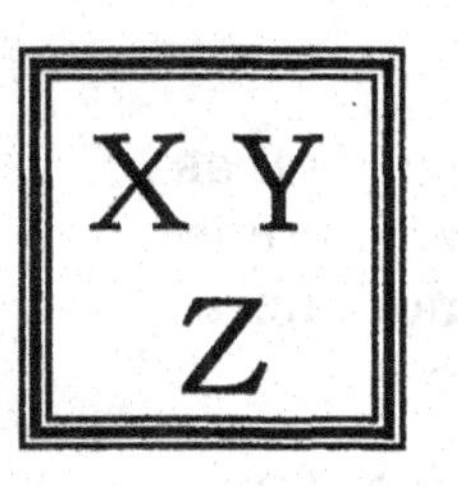

UNIT 7
Ratios and Proportional Reasoning

Section 1 Introduction

Section 2 Ratio and Direct Proportion

Section 3 Inverse Proportion

Section 4 Application Problems Involving Variation

Section 7-1: Introduction

Ratios and **proportions** are used in many situations in everyday life as well as in business and science. A common everyday example is cooking. If we increase or decrease a recipe, we have to keep ratios of the ingredients the same. There are many other instances where you have used ratios. The ratio of the circumference of a circle to the diameter is π (pi): $\dfrac{C}{d} = \pi$. The probability of an event occurring is a ratio. Probabilities are part-to-whole ratios: the number of favorable outcomes to the total number of possible outcomes. The probability of throwing a die and getting a 5 is 1 to 6, one favorable outcome to a total of six possible outcomes. When you read distances on a map, or measurements on a blueprint, you are using ratios.

The concept of ratio leads directly to the concept of proportional reasoning and proportions. Two equal ratios form a proportion. You can use proportions to find a quantity that you cannot easily measure. For example, you can figure out the height of a tree without measuring it by using a proportion.

Section 7-2: Ratio and Direct Proportion

In any field, size comparisons are often necessary. A useful method of comparing things is by a **ratio**, which is a concise way to express the relative sizes of two measures. The ratio of A to B means the quotient $\dfrac{A}{B}$. In 2009 there were 3,206 females and 2,511 males at Keene State College. The ratio of females to males is therefore, $\dfrac{3206 \; females}{2511 \; males}$. This ratio could also be written as $\dfrac{1.277 \; females}{1 \; male}$ or just 1.277, which tells us there are about 27.7% more females than males attending KSC.

Example 7-2.1: In 2005 the enrollment at Keene State College was 5,127 students and in 2009 it was 5,717 students. Find the percent increase from 2005 to 2009.

Solution: We previously solved this type problem using the formula

$$percent \; change = 100\left(\frac{new \; value - old \; value}{old \; value}\right).$$

We can also use ratios to find the percent change.

The ratio of student enrollment in 2009 to enrollment in 2005 is

$$\frac{5717}{5127} = 1.115.$$

This tells us that the enrollment grew by a factor of 1.115 or 11.5%. ♦

If an amount changes by a factor of 1.115, why does that mean it increased by 11.5 percent? Let A be the original amount, then

$$1.115A = (1 + 0.115)A = A + 0.115A.$$

You can see from this last equation that A increased by 0.115 which is 11.5%.

Activity 7.1

Use ratios to show that if you had $45,000 invested in a stock in 2008 and in 2009 the amount decreased to $38,500, that the percent decrease is about 14%.

Most people first encounter linearity via the concept of ratio, in a problem such as the following:

> *A soft drink recipe calls for mixing five cups of ginger-ale*
> *with two cups of lemon-lime soda. How much lemon-lime*
> *soda should you use to make 28 cups of the soft drink?*

A ratio describes a specific relationship between quantities. If the ratio of ginger-ale to lemon-lime soda is 5 to 2, then it means that for every 5 cups of ginger-ale you use, you must use 2 cups of lemon-lime soda, as shown in the figure below. You could also say that for every 2 cups of lemon-lime soda you use, you must use 5 cups of ginger-ale.

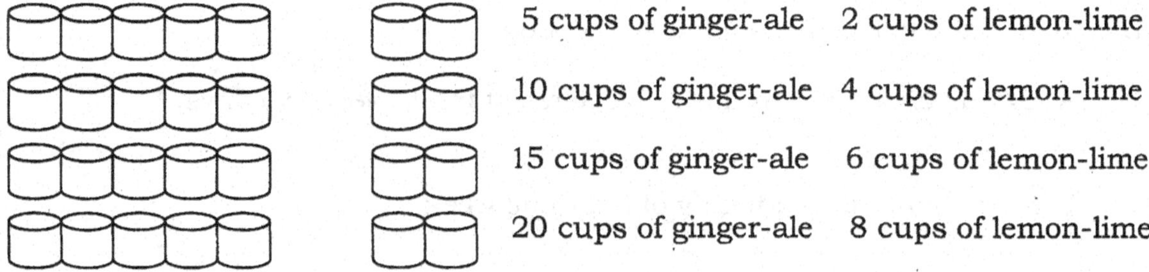

You can see from the figure that the answer to the question above is that you would use 8 cups of lemon-lime soda to make 28 cups of soft drink.

Knowing that the ratio of the amount of ginger-ale to the amount of lemon-lime soda is 5 to 2 we can write the following linear relationship:

$$G = \frac{5}{2}L,$$

where G represents the amount of ginger-ale and L represents the amount of lemon-lime soda. This is a linear function with slope $\frac{5}{2}$ and vertical intercept 0.

You could use Excel to solve the last problem by making a table such as the following:

	A Amount of lemon-lime (cups)	B Amount of ginger-ale (cups)	C Total (cups)
1			
2	0	0	0
3	1	2.5	3.5
4	2	5	7
5	3	7.5	10.5
6	4	10	14
7	5	12.5	17.5
8	6	15	21
9	7	17.5	24.5
10	8	20	28
11	9	22.5	31.5
12	10	25	35

Notice in any row the ratio of the amount of ginger-ale to the amount of lemon-lime is $\frac{5}{2}$. For example, in row 10, the ratio is 20 to 8, which is written as $\frac{20}{8}$ and $\frac{20}{8} = \frac{5}{2}$.

Activity 7.2

Write each of the following as a linear function.

a)   A logging crew can cut down five acres of trees every two days.

b)   It costs \$90 to feed a family of 3 for one week.

c)   The average human heart beats at 72 beats per minute.

Example 7-2.2: Algebraically solve the following problem,

A soft drink recipe calls for mixing five cups of ginger-ale with two cups of lemon-lime soda. How much lemon-lime soda should you use to make 28 cups of the soft drink?

Solution: Since the ratio of ginger-ale to lemon-lime soda is 5 to 2, we can write the linear function $G = \dfrac{5}{2}L$. The total amount of soft drink is

$$G + L = \text{total}.$$

Now substituting $\dfrac{5}{2}L$ for G, we get $\dfrac{5}{2}L + L = total$.

We are looking for the amount of Lemon-lime soda needed to make 28 cups of soft drink, therefore the equation we have to solve is

$$\frac{5}{2}L + L = 28$$

Solving this equation we get,

$$\frac{7}{2}L = 28$$

$$L = \frac{2}{7} \bullet 28 = 8$$

So it takes 8 cups of lemon-lime soda for 28 cups of soft drink. ♦

Activity 7.3

For a certain shade of paint, the ratio of white paint to blue paint is 3 to 2. In the table below fill in different amounts of blue and white paint that you would mix to get the same shade of paint.

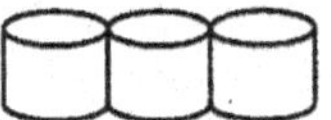 

Use ___ cups of white paint for ___ cups of blue paint to make ___ cups of paint.

Use ___ cups of white paint for ___ cups of blue paint to make ___ cups of paint.

Use 1 cup of white paint for ___ cups of blue paint to make ___ cups of paint.

Use ___ cups of white paint for ___ cups of blue paint to make 2 cups of paint.

In the last activity, suppose that to make 1 batch of paint we need 3 cups of white paint and 2 cups of blue paint. It does not matter how many batches of paint we make, the ratio of white to blue paint is still 3 to 2. If we made three batches of paint the amount of white and blue paint describe a different ratio, 9 to 6, but the two ratios 3 to 2 and 9 to 6 are equivalent. We can write this as the following equation:

$$\frac{3}{2} = \frac{9}{6}.$$

The statement that two ratios are equivalent is called a **proportion.** The two ratios are equivalent when the cross products are equal, i.e. $\frac{a}{b} = \frac{c}{d}$ if and only if $ad = cb$ provided $b \neq 0$ and $d \neq 0$.

We use the term **proportional** to indicate that one thing is a multiple of another. For example, when converting feet to inches we use

$$inches = 12 \times (number\ of\ feet).$$

Another way of expressing this is to say that the number of inches is **proportional** to the number of feet. If we let I denote the number of inches and F denote the number of feet then we have

$$I = 12F.$$

The symbol α is used to denote a proportionality relation. So $I\ \alpha\ F$ is shorthand for the phrase "I is proportional to F." In this example, 12 is called the **constant of proportionality**. It is a good idea to make sure the units fit. The constant of proportionality is 12 inches per foot. The units on the left side of the equation are inches and the right side is the product of inches per foot and feet as shown below. Treat the units as if they were algebraic quantities. As algebraic quantities they can be multiplied and then divided out.

$$I(inches) = 12(\frac{inches}{foot})F(feet)$$

$$inches = \frac{inches}{foot} \times feet$$

$$inches = inches$$

The key points from the last example are that proportional relationships are examples of linear functions of the form $f(x) = mx$, in other words, linear functions without a constant term and the units need to be consistent.

Example 7-2.3: Suppose you were driving at a constant rate of 40 miles per hour, then the distance you travel is proportional to the time you drive. If t is the time you drive, the distance you drive is $D(t)$. We know that $D(t)$ is the linear function: $D(t) = 40t$ and the constant of proportionality is 40.

◆

Activity 7.4

a) Show that the formula

$$F = 12I,$$

where I represents the number of inches and F denote the number of feet, could not be correct by checking units.

b) Suppose you have a job that pays \$9.00 per hour regardless of the number of hours you work. Write an equation that shows the proportionality relation. What is the constant of proportionality?

Example 7-2.4: A soft drink recipe calls for mixing five cups of ginger-ale with two cups of lemon-lime soda. How much ginger-ale should be mixed with 10 cups of lemon-lime soda?

Solution: Let's look at three methods of doing this problem.

Method 1

We can use proportional reasoning to determine how much ginger-ale we need. Since 10 cups of lemon-lime soda is 5 times as much as 2 cups, we should mix 5 times as much ginger-ale, $5 \times 5 = 25$ cups. The reasoning is summarized in the figure below.

	Problem		**1st step**		**2nd step**	
cups of ginger-ale	5	?	5	?	$5 \xrightarrow{\times 5} 25$	
cups of lemon-lime soda	2	10	$2 \xrightarrow{\times 5} 10$		$2 \xrightarrow{\times 5} 10$	

Method 2

We can set up a proportion and solve for the amount of ginger-ale
we need. Let N represent the amount of ginger-ale that has to be
added, so the ratio of ginger-ale to lemon-lime soda in the new

mixture is N to 10. Thus we have the proportion $\dfrac{5}{2} = \dfrac{N}{10}$.

Solving this equation, we get

$$\frac{5}{2} = \frac{N}{10}$$

$$2N = 5 \bullet 10 \qquad \text{(cross multiply ac = bd)}$$

$$N = 25 \qquad \text{(divide both sides by 2)}$$

Therefore, we will need 25 cups of ginger-ale.

Method 3

We can use the function notation we developed earlier:

$$G = \frac{5}{2}L ,$$

where G represents the amount of ginger-ale and L represents the
amount of lemon-lime soda. Since the questions asks how much
ginger-ale we need with 10 cups of lemon-lime soda we can
substitute 10 for L in the function and get

$$G = \frac{5}{2} \bullet 10 = 25 \ \text{ cups.} \qquad \qquad \blacklozenge$$

Example 7-2.5: A map has a scale of 2 inches for every 20 miles. If two
cities are 18 inches apart on the map, what is the actual
distance between them?

Solution:

Method 1

Since 18 inches is 9 times as much as 2 inches, the actual
distance should be 9 times 20 miles, or 180 miles.

Method 2

Let's set up the following proportion.

$$\frac{2}{20} = \frac{18}{D}\ ,\ \text{where } D \text{ represents the actual distance in miles.}$$

Solving the proportion $\dfrac{2}{20} = \dfrac{18}{D}$, we get

$$2D = 20 \bullet 18$$

$$D = \frac{20 \bullet 18}{2} = 180 \text{ miles}$$

Method 3

We can write the function

$$D = \frac{20}{2}I\ ,$$

where D represents the actual distance in miles and I represents the number of inches measured on the map. Since $I = 18$, we have

$$D = \frac{20}{2} \cdot 18 = 180 \text{ miles} \qquad\qquad\blacklozenge$$

Example 7-2.6: The ratio of the number of students in Joe's class to the number in Eileen's class was 3 to 4. The ratio became 9 to 10 when Joe added another 6 students. How many students did Joe have at first?

Solution: Let J represent the number of students originally in Joe's class and let E represent the number in Eileen's class. If the ratio is 3:4,

we can write $\qquad\qquad J = \dfrac{3}{4}E\ .$

Six students are added to Joe's class and the ratio becomes 9:10, which can be written as

$$J + 6 = \frac{9}{10}E$$

Substituting $\dfrac{3}{4}E$ for J into the second equation, we get

$$\frac{3}{4}E + 6 = \frac{9}{10}E\ .$$

Solving for E,

$$6 = \frac{9}{10}E - \frac{3}{4}E \qquad \text{(subtract } \frac{3}{4}E \text{ from both sides)}$$

$$6 = \frac{6}{40}E \qquad \text{(subtract fractions as in unit 2)}$$

$$E = \frac{240}{6} = 40 \qquad \text{(multiply both sides by } \frac{40}{6}\text{)}$$

Eileen has 40 students, so substituting 40 for E into the equation $J = \frac{3}{4}E$, we get Joe originally had 30 students. ◆

Activity 7.5

a) Suppose I can paint at a constant rate of k square feet per hour, if I can paint 400 square feet in 4 hours, how many square feet can I paint in 6 hours?

b) On a map 1 inch represents 15 miles. Find the actual distance of two towns that are 3.5 inches apart on the map.

At the time of this writing the exchange rate between the US dollar and the British pound was about $1.46 to the pound. The following table shows corresponding pairs of values using p and d to stand for the number of pounds and dollars respectively.

Pounds (p)	Dollars (d)
1	1.46
2	2.92
3	4.38
4	5.84
5	7.30

The two sets of numbers are proportional (directly proportional) because they are linked by the exchange rate. We can write the relationship between dollars and pounds as follows:

$$d = 1.46p$$

Graphing this relationship we get:

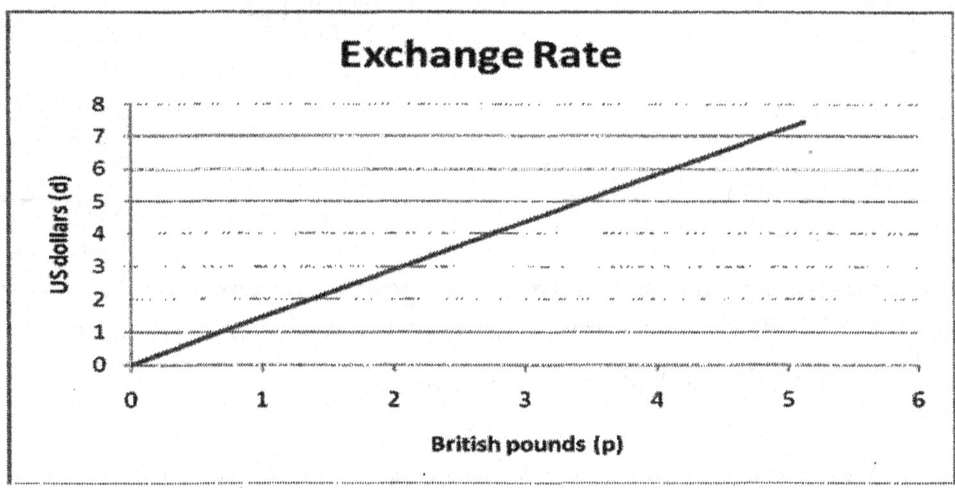

Notice the graph is a straight line with a positive slope that passes through the origin.

In the example above, we were given the value of the constant of proportionality and created a table showing the relationship between the English pound and the US dollar. Sometimes, we will be given data and asked if it is proportional. In order to determine if data is proportional, we need to recall that a direct proportion is written symbolically as $y \propto x$, which means that $y = kx$ where k is the proportionality constant. Solving this equation for k gives us $k = \dfrac{y}{x}$. For the data to be proportional, the ratio $\dfrac{y}{x}$ should be equal to a constant.

<u>**Example 7-2.7**</u>:   Given the following table, is $p \, \alpha \, F$?

F	26	31	36
p	91	108.5	126

Solution: If $p \, \alpha \, F$ then $p = kF$. So if p is proportional to F, $\dfrac{p}{F}$ should be equal to a constant. Checking this we get

$$\frac{91}{26} = 3.5, \quad \frac{108.5}{31} = 3.5, \quad \frac{126}{36} = 3.5$$

so $p \, \alpha \, F$. The constant of proportionality is 3.5.   ♦

Activity 7.6

Given the following table, are L and g directly proportional?

L	7.5708	18.9270	26.4978	45.4248	56.7810
g	2	5	7	12	15

Sometimes a quantity y varies directly as the power of a variable. For example, if $y \, \alpha \, x^3$ then $y = kx^3$. If you were given data, you could check this by examining the ratio $\dfrac{y}{x^3}$, it should be equal to a constant.

<u>**Example 7-2.8**</u>:   Given the following table, is $y \, \alpha \, x^2$?

x	2	4	6	8	10
y	8	32	72	128	200

Solution: If $y \, \alpha \, x^2$ then $y = kx^2$. So if y is proportional to x^2, $\dfrac{y}{x^2}$ should be equal to a constant. Checking this we get

$$\frac{8}{2^2} = 2, \quad \frac{32}{4^2} = 2, \quad \frac{72}{6^2} = 2, \quad \frac{128}{8^2} = 2, \quad \frac{200}{10^2} = 2$$

Hence, $y \, \alpha \, x^2$. The constant of proportionality is 2.   ♦

Activity 7.7

Given the following table, is $y \alpha x^3$?

x	1	4	6
y	4	256	864

Activity 7.8

A friend of yours looks at the following table and concludes that $y \alpha x$ because as x increases, so does y.

x	y
1	2
2	8
3	18
4	32
5	50

Is this conclusion correct? Why or why not?

If you plotted the points in the last activity you would get the following graph.

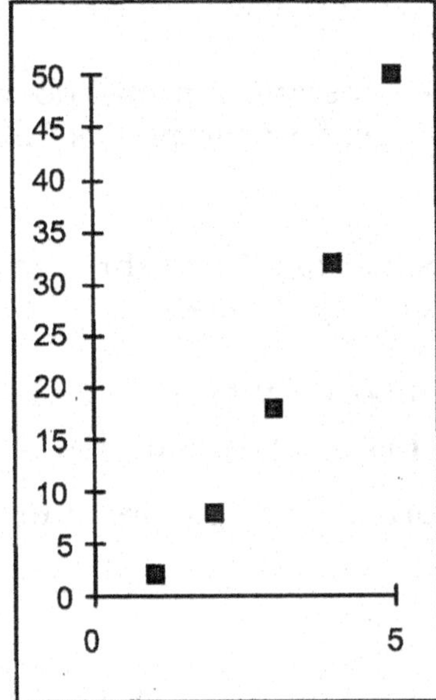

Notice the points do not lie on a line that passes through the origin. We said previously that if $y \alpha x$ "the points plot as a straight line through the origin."

Use the graph below to determine whether y varies directly as x.

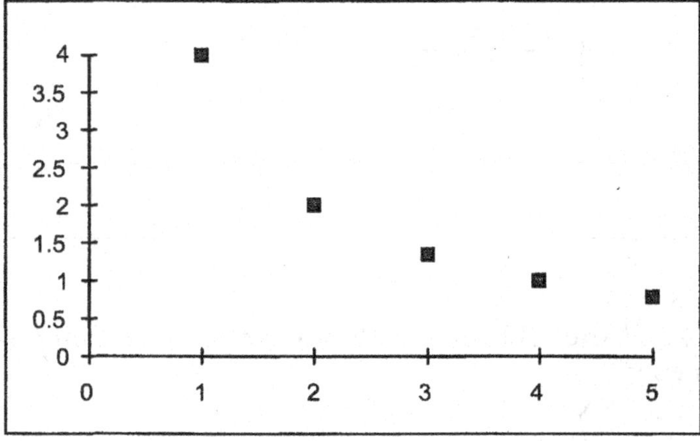

Remember two quantities are proportional when one is a multiple of the other. There are some key points to be made when two sets of numbers are proportional:

- As one variable doubles, triples, halves, etc., then so does the other.

- When one variable is zero then so is the other.

- Corresponding pairs of numbers are in the same ratio, which means that there is a single multiple linking each pair.

- The multiple, referred to as the constant of proportionality has different interpretations as a rate of change, for example, exchange rate, cost per unit, and speed.

- Plotting the two variables gives a straight-line through the origin whose slope is equal to the constant of proportionality.

- Direct proportion is written symbolically as $y \, \alpha \, x$, which means that $y = kx$ where k is a constant. This defining equation has the equivalent form $k = \dfrac{y}{x}$, which tells that corresponding pairs of numbers are in the same ratio.

Homework Section 7-2

1. In 2000 the population of New Hampshire was 1,235,786 people and in 2005 the population was 1,309,940. Use ratios to find the percent increase in the population.

2. The table below shows the breakdown of students at Keene State College by class rank.

Class Level	Fall 2009
Freshman	1637
Sophomore	1240
Junior	1156
Senior	938
Total	4971

    a) Find the ratio of seniors to freshman.

    b) How many percent fewer seniors are there than freshman?

3. If you mix 3 cups of blue paint with 5 cups of yellow paint to make a green paint, then how many cups of yellow paint do you need if you want to make the same shade of green paint using 4 cups of blue paint?

4. If you mix 3 cups of blue paint with 5 cups of yellow paint to make a green paint, write the amount of blue paint as a linear function of the amount of green paint that will make the same shade of green paint.

5. If you mix 3 cups of blue paint with 5 cups of yellow paint to make a green paint, how much yellow paint do you need to make 24 cups of this particular shade of green?

6. A recipe that serves 6 people calls for $2\frac{1}{2}$ cups of flour. How much flour will you need to serve 8 people, assuming that the ratio of people to cups of flour remain the same?

7. Given the following table, is $p\,\alpha\,F$

F	15	18	21
p	37.5	45	52.5

If $p\,\alpha\,F$, what is the constant of proportionality? Write p as a linear function of F.

8. A 40-pound child should receive how many units of penicillin if the dosage is 500,000 units for a 150-pound adult?

9. The ratio of the area of the dining room to the family room in Brian's house is 2 to 3. After remodeling, the family room is now $\frac{1}{2}$ as large as it used to be and has 60 sq feet less than the dining room. How many square feet is the dining room?

<u>**Section 7-3: Inverse Proportion**</u>

In example 2.3 we saw that if you were driving at a constant rate of 40 miles per hour, then the distance you travel is proportional to the time you drive. Some quantities though are related in a different way. As one quantity increases the other quantity decreases. This kind of relationship is called an **inverse proportion**.

Let's again use the formula that relates distance traveled, rate of travel, and time of travel but in a different form:

$$Time = \frac{Distance}{Rate} \text{ or } T = \frac{D}{R}$$

Suppose you had to drive 120 miles along a straight highway at a constant speed, how does T depend on R? To help answer this question examine the following table.

D	R	$T = \dfrac{D}{R}$
120 miles	10 mph	12 hr
120 miles	20 mph	6 hr
120 miles	30 mph	4 hr
120 miles	40 mph	3 hr
120 miles	50 mph	2.4 hr

Examining the table you can see that if you *double* the speed, it takes *one-half* as long, if you *triple* the speed, it takes *one-third* as long, and so on. Increasing the speed decreases the time, for any fixed distance. Notice that if you multiply T times R you get a fixed constant, the distance traveled. We say that "R is **inversely proportional** to T" or "R **varies inversely** as T." Symbolically, $R \alpha \dfrac{1}{T}$, this would be written as $R = \dfrac{k}{T}$, where k is the constant of proportionality. In this example, k is equal to 120.

In general we can say that:

A quantity y is inversely proportional to x if $y = \dfrac{k}{x}$, where k is a constant. k is referred to as the proportionality constant. In other words, two variables x and y **are inversely proportional** when $x \cdot y$ is **constant**.

Determine whether the following variables are directly or inversely proportional.

a)

x	5	10	15	20	25
y	2	4	6	8	10

b)

x	5	10	15	20	25
y	24	12	8	6	4.8

The table below shows three examples of directly and inversely proportional variables written verbally, symbolically, and as formulas.

Verbal	Symbolical	Formula
The length, L, of the skid marks of a car's tires (when the brakes are applied) varies directly as the square of the velocity, v, of the car.	$L \propto v^2$	$L = kv^2$
The pressure, P, varies directly as the temperature, T.	$P \propto T$	$P = kT$
At a given temperature, the pressure p of an ideal gas varies inversely as the volume v. (Boyle's Law)	$p \propto \dfrac{1}{v}$	$p = \dfrac{k}{v}$

Activity 7.11

Write the following verbal statements as formulas.

a)   A varies directly as the square of B.

b)   V varies directly as the cube of e.

c)   y varies inversely as the cube of x.

d)   C varies inversely as D.

Example 7-3.1: An oceanographer took readings of the water temperature C (in degrees Celsius) at depth d (in meters). The data is given in the following table.

d	1000	2000	3000	4000	5000
C	4.2	1.9	1.4	1.2	0.9

How can we tell if C varies inversely as d, in other words, does $C = \dfrac{k}{d}$?

Solution: If C and d **vary inversely** then C·d is **constant.** Looking at the data we have

$$4.2(1000) = 4200$$
$$1.9(2000) = 3800$$
$$1.4(3000) = 4200$$
$$1.2(4000) = 4800$$
$$0.9(5000) = 4500$$

Since the products are <u>not constant</u>, then C is not inversely proportional to d. When collecting real data you cannot expect to get the products exactly equal, but some of these vary a little too much.   ◆

Remember you can think of inverse proportions in two ways:

- $y = \dfrac{k}{x}$

- $xy = k$

Homework Section 7-3

1.   Determine whether the following variables are directly or inversely proportional.

a)

x	4	6	8	10
y	12	18	24	30

b)

p	2	4	5	8
F	20	10	8	5

c)

r	4	7	10	15
v	12	8	5	3

2.   Suppose the winning ticket in a lottery pays $1,000,000, complete the table below.

Number of Winners	Amount Each Winner Receives
1	$1,000,000
2	
	$200,000
10	
	$40,000

Write a formula that relates number of winners to the amount each winner receives.

3.   Suppose you want to fence in 2,000 square feet of your backyard to make a rectangular play area. We know the formula for the area of a rectangle is *length x width*. Since we want $length \times width = 2000$, we can say that length and width vary inversely. This could also be written as $length = \dfrac{2000}{width}$. If the length of the play area was 80 feet, how long would the width be?

In this section we look at more applications of variation.

Example 7-4.1: A company has found that the demand for its product varies inversely as the price of the product. When the price is $3.75, the demand is 300 units. Approximate the demand when the price is $4.50.

Solution: Since d varies inversely as p, we know that $d = \dfrac{k}{p}$. This could also be written as $k = dp$. So we have $d_1 = 300$, $p_1 = 3.75$ and we have to find d_2 when $p_2 = 4.50$. Since $d_1 p_1 = k$ and $d_2 p_2 = k$, we can say that $d_1 p_1 = d_2 p_2$. Substituting, we get

$$300(3.75) = 4.50 d_2 .$$

Solving for d_2, we get

$$d_2 = \frac{300(3.75)}{4.50} = 250.$$

Hence when p = $4.50, the demand is approximately 250 units.

♦

Example 7-4.2: Your weight on the moon is directly proportional to your weight on the Earth. Suppose you weigh 135 lbs on Earth, then on the moon you would weigh 22.5 lbs. How much would your friend weigh on the moon if he weighs 174 lbs on Earth?

Solution: Since W_m varies directly as W_e, we know that $W_m = kW_e$. This could also be written as $k = \dfrac{W_m}{W_e}$, which says that the ratio of W_m to W_e is constant. Putting the data we in a table, we have

W_e	135	174
W_m	22.5	x

↑   ↑
your friend's
weight weight

Since we know the ratios have to be constant, we can write

$$\frac{135}{22.5} = \frac{174}{x} .$$

Cross multiplying and solving for x, we get $x = \dfrac{(22.5)(174)}{135} = 29$.

Therefore, your friend will weigh 29 lbs on the moon if he weighs 174 lbs on Earth.   ◆

Activity 7.12

Solve the following.

a)   y varies inversely with the square of x. If y = 4 when x = 5, find y when x = 10.

b)   The cost C, of producing x calculators varies directly as x. If it costs $4050 to produce 75 calculators, what would the cost be to produce 250 calculators?

Unit 7 Review Exercises

1. Match each description from column 1 with a possible formula from column 2. Each item in column 1 **may have more than one** possible match from column 2.

<u>Column 1</u>

a) y varies directly as x

b) $y \propto x^2$

c) y varies inversely as the cube of x

d) y varies directly as the square root of x

<u>Column 2</u>

1. $y = 2x + 4$

2. $y = 4\sqrt{x}$

3. $y = 10x$

4. $y = kx^2$

5. $y = \dfrac{6}{x^3}$

6. $y = kx^3$

2. Given the table below, is $r \propto t$? Explain.

t	1	2	3	4
r	6	12	18	24

3. Set up the formula from the given statements.

a) r varies directly as t.

   Solution:
   $$r = kt$$

b) p varies inversely as the square of q.

c) w is proportional to the cube of L.

d) y is inversely proportional to the square of x and is directly proportional to W.

4.   Express the meaning of the given equation in a verbal statement.

$$f = k\sqrt{n}$$

5.   a)   If r is inversely proportional to d^2 and r = 2 when d = 4, find r if
                d = 8.

        b)   Find the constant of proportionality, k, if a varies directly as the
                square of b and a = 20 when b = 2.

        c)   If y varies directly as x and y = 20 when x = 8, find y when x = 10.

        d)   If z varies directly as the square of x and inversely as y, and z = 6
                when x = 6 and y = 4, find z when x = 3 and y = 4.

        e)   The time it takes a pendulum to make one swing (back and forth) is
                directly proportional to the square root of its length. A 1-foot
                pendulum takes 6 seconds to complete one swing. How long does it
                take a 3-foot pendulum to complete one swing?

6.   The diameter of the largest particle that can be moved by a stream varies
        directly as the square of the velocity of the stream. A stream with a velocity
        of 0.25 miles per hour can move course sand particles about 0.02 inch in
        diameter. Approximate the velocity required to carry particles 0.12 inch in
        diameter.

7.   The volume of a gas is inversely proportional to the pressure. If a
        pressure of 36 pounds per square inch corresponds to a volume of 25
        cubic feet, what pressure is needed to produce a volume of 75 cubic
        feet?

<u>**Section 8-1: Introduction**</u>

Quantities have many units, which can be used to measure them. The following table gives common units associated with certain dimensions.

Dimension	Common units
Time	seconds, minutes, hours days, months, years
Distance	inches, feet, yards centimeters, meters
Speed	miles per hour, feet per second meters per second
Volume	cubic feet, cubic yards cubic meters, liters

When using formulas it is often necessary to change units. For instance, if you had to find the distance traveled using the formula $d = vt$, and $v = 30$ miles per hour and $t = 45$ minutes, you would have to convert time to hours or velocity to miles per minute because the units are not the same.

<u>**Section 8-2: Unit Conversions**</u>

In order to convert units, we will first write the proportional relationship between the units. For example, suppose we wanted to convert 2 feet to inches. To do this we use the conversion 1 foot is equal to 12 inches and writing this relationship algebraically we get

$$I = 12F, \text{ where } I \text{ is the number of inches and } F \text{ is the number}$$
$$\text{of feet.}$$

At first this function may look backwards, but it is saying that the number of inches is equal to the number of feet times 12. It is **not** saying that one inch is equal to 12 feet. If you look at a table below it will help verify the formula.

Feet	Inches
1	12
2	24
3	36
4	48

It is important to understand that in the formula such as $I = 12F$, the formula has to make sense dimensionally. In other words, units have to be consistent. In the formula above, 12 is a constant, but there are units attached to it.

We really have $\dfrac{12\ in}{1\ ft}$ or $12\dfrac{in}{ft}$.

If you just focus on the units you can see that the units are consistent on both sides of the equation.

$$inches = \frac{inches}{feet} \times feet$$

You can see that the units act as numbers, and the feet divide out, leaving us with *inches = inches*.

Below is a table giving come common conversions.

Common Conversions
8 fluid ounces = 1 cup
2 cups = 1 pint
2 pints = 1 quart
4 quarts = 1 gallon
8 pints = 1 gallon
3 teaspoons = 1 tablespoon
16 tablespoons = 1 cup
16 fluid ounces = 1 pint
16 ounces = 1 pound
5,280 feet = 1 mile
12 inches = 1 foot
3 feet = 1 yard

Example 8-2.1: Write the proportional relationship relating pints to gallons.

Solution: The conversion to use is 1 gallon is equal to 8 pints, so

$$P = 8G \text{, where } P \text{ is the number of pints and } G \text{ is the}$$
$$\text{number of gallons} \qquad \blacklozenge$$

Check the units in the last example to make sure that the equation is dimensionally correct.

Activity 8.1

a)   Write the proportional relationships relating

   i)   feet to miles   ii)   feet to yards

b)   Using parts i) and ii) above, write a relationship between miles and yards.

Example 8-2.2: Convert 2.1 miles to yards

Solution: From the last activity, we have F = 5280M and F = 3Y, where F is equal to the number of feet, M is equal to the number of miles, and Y is equal to the number of yards.

Using these proportions, we have

$$F = 5280(2.1)$$

$$5280(2.1) = 3Y$$

Solving for Y, we get

$$Y = 3696 \text{ yards}$$

So, 2.1 miles = 3696 yards. ◆

Example 8-2.3: Convert 3.5 liters (L) to quarts (qt).
(1 L = 1.057 qt)

Solution: We have the relation, qt = 1.057L, where qt is the number of quarts and L is the number of liters.

So, 1.057(3.5) = 3.69 quarts

3.5 liters = 3.69 quarts ◆

Example 8-2.4: If I can swim 20 yards in 15 seconds, what is my speed in miles per hour?

Solution: Twenty yards in 15 seconds means 20 yards per 15 seconds which can be written as $\dfrac{20 \; yards}{15 \sec}$. We have to convert this so our units are $\dfrac{miles}{hour}$. In other words, yards have to be converted to miles and seconds to hours. Initially it might be easier to do one at a time, so start with converting yards to miles. We need the following conversions:

1 yard = 3 feet	and	1 mile = 5280 feet

The conversions on the previous page give us the following proportional relationships:

$$F = 3Y \text{ and } F = 5280M$$

Combining these, we get

$$5280M = 3Y \text{ or}$$

$$M = \frac{3}{5280}Y$$

So, 20 yards = 60/5280 miles

Now, $\dfrac{20\,yards}{15\,sec} = \dfrac{60/5280\,miles}{15\,sec}$

To finish, we have to convert 15 seconds to hours.

S = 60M and M = 60H, where S is the number of seconds, M is the number of minutes and H is the number of hours

Combining these, we get

$$S = 60(60)H$$

If S = 15, then H = 15/(60*60)

Therefore, $\dfrac{20\,yards}{15\,sec} = \dfrac{60/5280\,miles}{15/(60*60)\,hr} = 2.73 \; miles\;per\;hour$ ♦

Another method to do the last example, which is a condensed version, is to concentrate on the conversion factors and getting units to divide out. The following illustrates this method.

$$\frac{20\,yards}{15\,sec} \times \frac{3\,feet}{1\,yd} \times \frac{1\,mile}{5280\,feet} \times \frac{60\,sec}{1\,min} \times \frac{60\,min}{1\,hr} = \frac{20 \times 3 \times 60 \times 60\,miles}{15 \times 5280\,hr} = 2.73 \text{ mph}$$

$$\uparrow \qquad \uparrow \qquad \uparrow \qquad \uparrow$$
$$1 \qquad 1 \qquad 1 \qquad 1$$

You can see that the above method works because we are repeatedly multiplying $\dfrac{20\,yards}{15\,sec}$ by 1. For example, 3 ft = 1 yd so $\dfrac{3\,feet}{1\,yd} = 1$.

 A 15-pound case of bacon cost \$18. On the average, there are 20 slices of bacon in a pound. If one serving consists of three slices of bacon, what is the cost per serving of bacon?

Solution: We first have to make the assumption that

$$\text{bacon cost } \alpha \text{ number of pounds of bacon.}$$

Now we can say that

$$C = kP \qquad \text{where C is bacon cost and}$$
$$\text{P is the number of pounds}$$

Since a 15-pound case costs \$18, we can write

$$18 = k(15)$$

$$\text{So, } k = \frac{18}{15} = \frac{6}{5}$$

Therefore, $C = \frac{6}{5}P$.

Since there are 20 slices per pound,

$$S = 20P, \text{ where S is the number of slices}$$

One serving consists of three slices, so

$$S = 3Se, \text{ where Se is the number of servings.}$$

Keep in mind what we are looking for the cost per serving or $\frac{C}{Se}$.

To find $\frac{C}{Se}$ let's use the three relationships we have found.

Using $S = 3Se$ and $S = 20P,$ we can say

$$3Se = 20P.$$

Solving for P, we get

$$P = \frac{3Se}{20}$$

Substituting $P = \dfrac{3Se}{20}$ into the first equation $C = \dfrac{6}{5}P$, we get

$$C = \frac{6}{5}\left(\frac{3}{20}Se\right).$$

Therefore, $\dfrac{C}{Se} = \dfrac{18}{100} = 0.18$ and the cost per serving is 18 cents.   ♦

Using the more compact form, we could write

$$\frac{18\ dollars}{15\ lb} \times \frac{1\ lb}{20\ slices} \times \frac{3\ slices}{1\ serving} = \frac{18 \times 3\ dollars}{15 \times 20\ serving} = 0.18\frac{dollars}{serving}$$

When using this compact form, you have to keep in mind what you are solving for. Here we wanted cost per serving, that is why we started with $\dfrac{18\ dollars}{15\ lb}$ instead of $\dfrac{15\ lb}{18\ dollars}$.

Activity 8.2

A 20-pound box of elbow macaroni costs \$12. One portion of cooked macaroni is about 10 ounces. However, cooked macaroni weighs about 2.6 times as much as the raw macaroni. Find the cost per portion.

Name:______________________________________   **Date:**____________

Homework Section 8-2

1. Using the following conversion table below write the proportional relationships

    a) between American dollars and Canadian dollars.

    b) between American dollars and Euros.

	United States dollar (USD)	British pound (GBP)	Canadian dollar (CAD)	European Euro (EUR)
USD	1	1.45027	0.943734	1.1959
GBP	0.689523	1	0.650726	0.8246
CAD	1.05962	1.53674	1	1.2672
EUR	0.83619	1.2127	0.789141	1

2. Use your answer from 1a) above to convert 6 American dollars to Canadian dollars.

3. Convert 2.5 British pounds to Euros.

4. A baseball pitcher throws a baseball 90 mph; convert this to feet per second.

5. If a pitcher can throw a baseball at 90 mph, how long does it take the ball to go from the pitcher's mound to home plate (60 feet 6 inches)?

You have to be careful in converting units that are raised to a power. For example, suppose you wanted to convert square yards (yd^2) to square feet (ft^2). We know that 1 yd = 3 ft, so we can find the conversion factor between square yards (yd^2) to square feet (ft^2) by writing:

$$1\ yd^2 = 1\ yd \times 1\ yd = 3\ ft \times 3\ ft = 9\ ft^2$$

That is,

$$1\ yd^2 = 9\ ft^2.$$

Note that we squared both sides of the yards-to-feet conversion:

$$1\ yd = 3\ ft \quad \xrightarrow{\text{Square both sides}} \quad (1\ yd)^2 = (3\ ft)^2 \quad or \quad 1\ yd^2 = 9\ ft^2$$

Example 8-3.1:   Convert $3\ ft^2$ to in^2.

Solution: We know 1 ft = 12 in, so squaring both sides we get $1\ ft^2 = 144\ in^2$. Therefore,

$$3\ ft^2 \times \frac{144\ in^2}{1\ ft^2} = 3 \times 144\ in^2 = 432\ in^2 . \qquad \blacklozenge$$

Activity 8.3

Convert $54\ ft^3$ to yd^3.

Hint: 1 yd = 3 ft, so cubing both sides we get $1\ yd^3 = 27\ ft^3$.

The metric system is very similar to our decimal number system. It is an easy system to use because calculations are based on multiples of 10. Different units of length in the metric system are obtained by combining an appropriate prefix with the base unit, which is the **meter**. The prefixes, their symbols, and their meanings are given in the following table.

Prefix	Symbol	Power of 10	Meaning
mega	M	10^6	one million times
kilo	k	10^3	one thousand times
hecto	h	10^2	one hundred times
deka	da	10^1	ten times
deci	d	10^{-1}	one tenth of
centi	c	10^{-2}	one hundredth of
milli	m	10^{-3}	one thousandth of
micro	μ	10^{-6}	one millionth of

Activity 8.4

Complete the following table, which gives names for different units of length along with their relationship to the meter.

Unit	Symbol	Relationship to Base Unit
kilometer	___	1000 meters
_____	___	100 meters
_____	dam	_______
meter	m	base unit
_____	dm	0.1 meter
centimeter	___	_______
_____	mm	_______

It is important to have some idea as to the size of some of these units. A **meter** is about 39 inches which is a little longer than a yard (36 inches). One **centimeter** is about the width of your little finger or the diameter of the head of a thumbtack. Nine football fields, including end zones, laid end to end are approximately 1 **kilometer** long. The thickness of a dime is about 1 **millimeter**.

Activity 8.5

Which metric unit (km, m, cm, or mm) should be used to measure each item?

a)   Length of a pencil   ____

b)   Your height   ____

c)   Length of a car race   ____

d)   Diameter of a nickel   ____

It is important to be able to change units so that you may work with them within a problem. Next, we will look at some examples that convert from one metric unit to another metric unit.

Example 8-4.1:   Change 2.3 km to meters.

Solution: Since kilo means 10^3 or 1000, **1 km = 1000 m**.

$$2.3 \text{ km} \times \frac{1000 \, m}{1 \, km} = 2300 \text{ m}$$

Example 8-4.2:   Change 5 meters to centimeters.

Solution: Since centi means 10^{-2} or 0.01, **1 cm = 10^{-2} m**.

$$5 \text{ m} \times \frac{1 \text{ cm}}{10^{-2} \, m} = 500 \text{ cm}$$

Some other metric conversion facts to keep in mind are given below.

$$1 \text{ m} = 100 \text{ cm} \quad 1 \text{ m} = 1000 \text{ mm} \quad 1 \text{ cm} = 10 \text{ mm}$$

Activity 8.6

a)   Change 125 mm to cm.

b)   Change 48 cm to km.

Another method used to make metric conversions is to use the following table.

*	*	*	*	*	*	*
kilo(k)	*hecto*(h)	*deka*(da)	meter	*deci*(d)	*centi*(c)	*milli*(m)

In the last example you had to change 125 mm to cm. Looking at the table above, to go from mm to cm you would have to move one place to the left, so to change 125 mm to cm, move the decimal point one place to the left.

$$125 \text{ mm} = 12.5 \text{ cm}$$

Example 8-4.3: Change 2.34 km to cm.

Solution: According to the table above, to go from km to cm you have to move 5 places to the right. Therefore to change 2.34 km to cm, you have to move the decimal point 5 places to the right.

$$2.34 \text{ km} = 234,000 \text{ cm}.$$

♦

Activity 8.7

Using the method in Example 8-4.3, convert the following.

a)   Change 6.14 mm to m.

b)   Change 0.75 km to m.

The **mass** of an object is the quantity of material making up the object. The base unit for mass in the metric system is the **gram**. A common paper clip has a mass of about 1 gram (g). One grain of salt has a mass of about 1 mg.

Activity 8.8

Which metric unit (kg, gram(g), cg, or mg) should be used to measure each item?

a)   Mass of a pencil   _____

b)   Mass of a football player _____

Example 8-4.4:   Change 5.8 kg to g.

Solution: We can use the same table as we did for length; we just need to
change the meter to the gram. The prefixes stay the same. To go
from kg to g, we need to move 3 places to the right. Therefore, to
change 5.8 kg to g, we have to move the decimal point 3 places to
the right to get

        5.8 kg = 5800 g.   ◆

Activity 8.9

a)   Change 120 grams to kg.

b)   Change 0.72 cg to mg.

Unit 8 Review Exercises

Use the following table of some common conversion factors to answer the following conversion problems.

1 in = 2.54 cm	1 lb = 453.6 g	1 ft^3 = 28.32 liters (L)
1 km = 0.6214 mile	1 kg = 2.205 lb	1 L = 1.057 qt
1 yd = 3 ft	1 lb = 4.448 N	
1 mi = 5,280 ft		

1. Convert 9.6 qt to liters.   2. Convert 170 lb to grams.

3. Convert $1 ft^2$ to square inches.   4. Convert $5 in^2$ to square cm.

5. Convert 12 km to miles.   6. Convert 32 in to cm.

7. Suppose I can swim 40 yards in 30 seconds, convert this to miles per hour.

Solution:

$$\frac{40\ yd}{30\ s} \times \frac{ft}{\_\_\ yd} \times \frac{mi}{\_\_\ ft} \times \frac{s}{\_\_\ min} \times \frac{min}{\_\_\ hr} = \frac{miles}{\_\_\ hour}$$

8. A foreign car's gasoline tank holds 56 L. Convert this capacity to gallons.

9. Twenty grams of a medication are to be dissolved in 0.060 L of water. Convert this to milligrams per deciliter.

10. The acceleration due to gravity is about 980 cm/s^2. Convert this to feet per square second.

11. Convert the following.

    a)   7.25 km to m   b)   0.108 mg to g

    c)   18 mm to cm   d)   25 kg to cg

12. Is the equation $x = x_0 + v_0 t + \frac{1}{2} at^2$ dimensionally correct?

    (a is acceleration (m/s^2), v_0 is velocity (m/s), t is time (s), and x and x_0 are lengths (m))

As we will see trigonometry has wide applicability which is remarkable for something so simple. The mathematical basis for trigonometry is **similar** triangles. Two figures are **similar** if they both have the same shape but not necessarily the same size. When a figure is put up on the overhead projector the image on the screen is similar to the image on the transparency. When you look at a map you are given a **scale factor,** which tells how its size compares to the actual size. The scale factor is also called the **ratio of similitude**.

There are two important properties of similar triangles.

> **Properties of Similar Triangles**
>
> 1. The corresponding angles of similar triangles are equal.
>
> 2. The corresponding sides of similar triangles are proportional.

If either of these properties is true for two triangles, the triangles are similar.

Example 9-1.1: Is the following pair of triangles similar? If they are similar, explain why and give the scale factor.

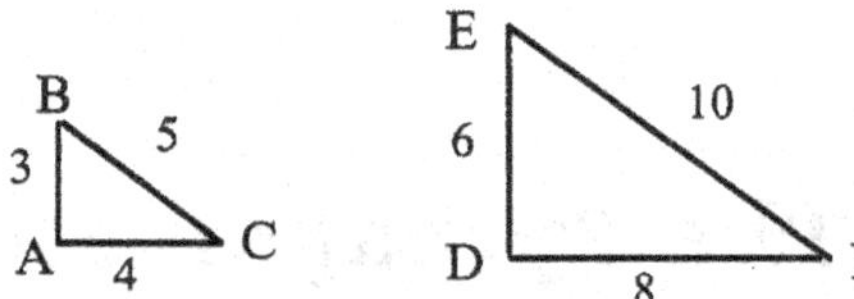

Solution: Yes, the triangles are similar because corresponding sides are proportional.

$$\frac{6}{3} = \frac{8}{4} = \frac{10}{5}$$

The scale factor (common ratio) is 2.   ♦

Example 9-1.2: Given that the following triangles are similar, find the missing length x.

Solution: Since the triangles are similar, corresponding sides are proportional.

Thus $\dfrac{2}{3} = \dfrac{10}{x}$. Solving the proportion we get $2x = 30$, so $x = 15$.   ♦

Consider the two **right** triangles below for which $\angle A = \angle A'$. In addition, we know that $\angle C$ and $\angle C'$ are both right angles (*measure 90°*). We can conclude that $\angle B$ has to equal $\angle B'$ since the sum of the angles in each triangle is 180°. Hence we conclude that if one acute angle (*angle less than 90°*) of a right triangle equals an acute angle of another, all three angles of the first triangle equal the corresponding angles of the other and the two triangles are **similar**. Since they are similar, the lengths of corresponding sides have the same ratio. That is,

$$\frac{A'B'}{AB} = \frac{B'C'}{BC} = \frac{A'C'}{AC}.$$

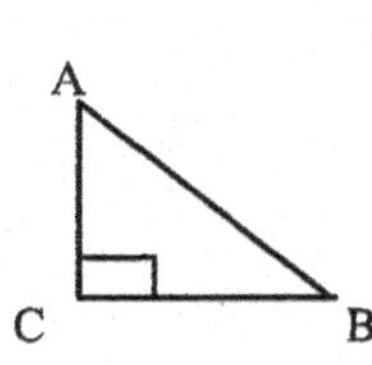 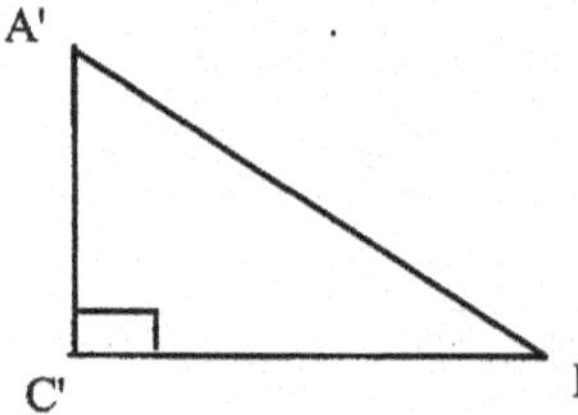

The proportions above are not quite in the form we want. Notice the ratios have lengths of sides from different triangles. Using algebraic manipulation we can get the ratios so the lengths of the sides are from the same triangle. In other words we can get proportions like the following:

$$\frac{BC}{AB} = \frac{B'C'}{A'B'}$$

To accomplish this you will need to follow the next example.

Example 9-1.3:   If $\dfrac{x'}{x} = \dfrac{y'}{y}$, where x, x', y, and y' $\neq$ 0, show that $\dfrac{y}{x} = \dfrac{y'}{x'}$.

Solution: If $\dfrac{x'}{x} = \dfrac{y'}{y}$ then cross products have to be equal.

So, $x'y = xy'$

$$\frac{x'y}{x} = y' \quad (\textit{dividing both sides by x})$$

$$\frac{y}{x} = \frac{y'}{x'} \quad (\textit{dividing both sides by x'}) \qquad \blacklozenge$$

Activity 9.1

From the proportion $\dfrac{A'B'}{AB} = \dfrac{B'C'}{BC} = \dfrac{A'C'}{AC}$, use algebra to show how the following proportions were found.

$$\frac{BC}{AB} = \frac{B'C'}{A'B'}$$

$$\frac{AC}{AB} = \frac{A'C'}{A'B'}$$

$$\frac{AC}{BC} = \frac{A'C'}{B'C'}$$

The ratios above supply the key to understanding right triangle trigonometry. They say that **if ΔA'B'C' is similar to ΔABC, the ratio of any two sides of ΔA'B'C' will equal the corresponding ratio in ΔABC.**

Example 9-1.4:   Given that $\angle A = \angle A'$, find x. $\angle C$ and $\angle C'$ are both right angles.

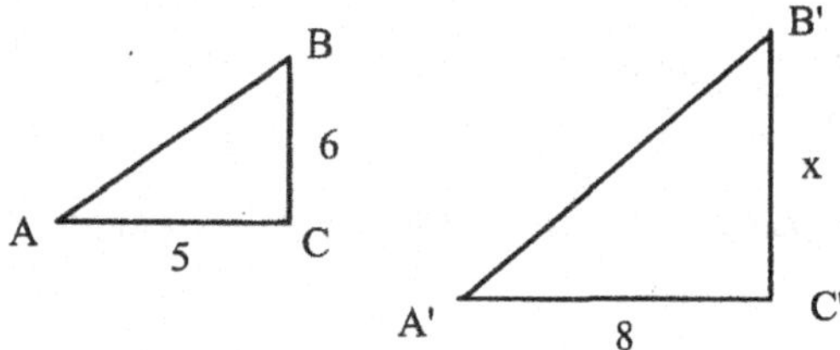

Solution: We know that ΔABC and ΔA'B'C' are similar since corresponding angles are equal. Therefore from the ratios above we can write

$$\frac{x}{8} = \frac{6}{5}.$$

Solving this equation for x by cross multiplying, we get

$$5x = 48$$

$$x = \frac{48}{5} = 9.6$$

So **x = 9.6** units.   ◆

Remember if $\triangle A'B'C'$ is **any** right triangle where $\angle A = \angle A'$, **the ratio of any two sides of that triangle will equal the corresponding ratio in $\triangle ABC$.** We have to consider than only the ratios that can be formed in $\triangle ABC$. Since these ratios are important, they are given special names. These ratios vary with the size of $\angle A$, so we are to speak of these ratios with reference to the angle. The three important trigonometric ratios are

$$\text{sine } A = \frac{BC}{AB} = \frac{\text{side opposite } \angle A}{\text{hypotenuse (\textit{side opposite the right angle})}}$$

$$\text{cosine } A = \frac{AC}{AB} = \frac{\text{side adjacent to } \angle A}{\text{hypotenuse}}$$

$$\text{tangent } A = \frac{BC}{AC} = \frac{\text{side opposite } \angle A}{\text{side adjacent to } \angle A}.$$

These trigonometric ratios are referred to by the shorter names sin A, cos A, and tan A respectively. It is imperative that you understand the ideas behind these ratios.

Consider two similar right triangles that have an acute angle of $40°$. One triangle is so small that we have to use a magnifying glass to see it and the other has sides that are measured in miles, the ratio of the side opposite the $40°$ angle to the side adjacent to the $40°$ angle is the same in both. This is the key to understanding how to work with these ratios. In addition, we can easily find this particular ratio using a calculator and entering tan $40°$.

In order to properly use these trigonometric ratios, the triangles have to be labeled correctly. This is covered in Section 2 on the next page.

It is very important that triangles are labeled correctly. The longest side of a right triangle is called the **hypotenuse** and is opposite the right angle. If we choose one of the acute angles, which we will call the **angle of interest**, the angle is made up of two sides of the triangle, the hypotenuse and the **adjacent side**. The remaining side is called the **opposite side** of the angle of interest.

Note: In any right triangle the hypotenuse does not change, but depending on what acute angle you are interested in, the opposite side and adjacent side change.

In the triangles below, notice that when $\angle A$ is the angle of interest, $\overline{BC}$ is the opposite side, but when $\angle B$ is the angle of interest, $\overline{AC}$ is the opposite side. Therefore, you must be very conscious of what angle is the angle of interest when setting up the ratios.

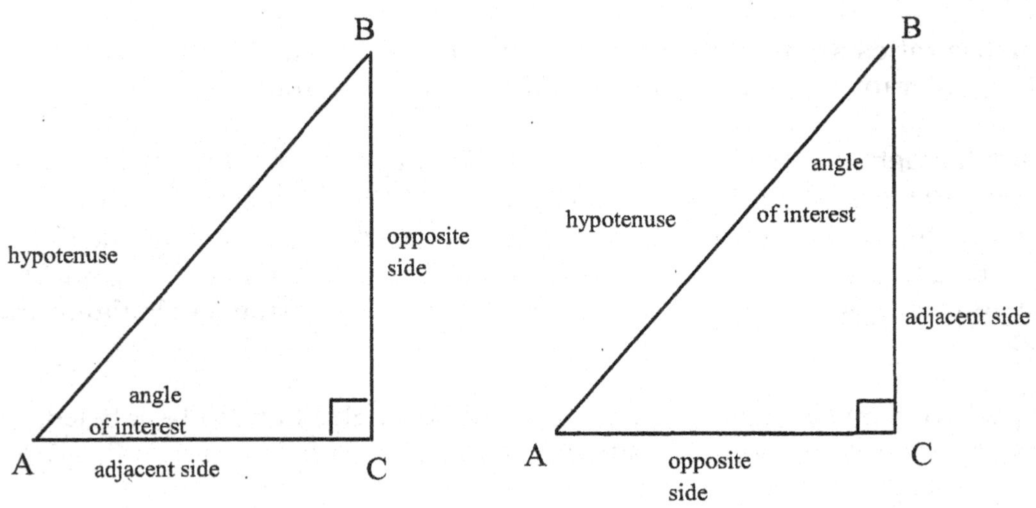

Note: From now on we will label the side opposite $\angle A$ as **a**, the side opposite $\angle B$ as **b**, and the hypotenuse (side opposite the right angle) as **c**.

Example 9-2.1: Given the triangle below, find the side adjacent to $\angle A$.

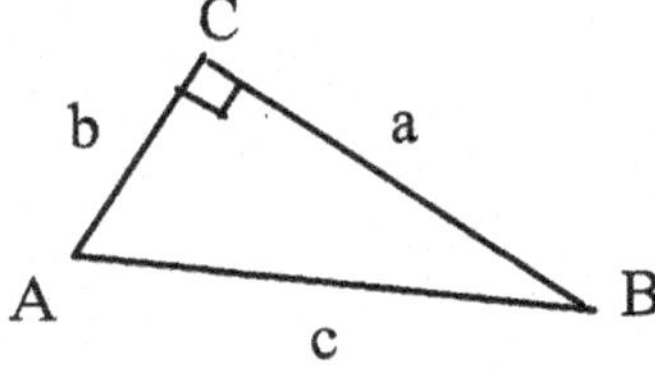

Solution: $\angle A$ is made up of two sides, the hypotenuse and the adjacent side b. So side b is adjacent to $\angle A$. ♦

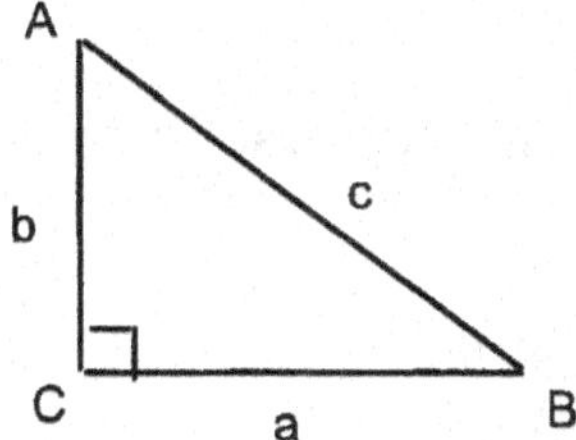

In the right triangle shown , what is
a) The side opposite angle A ?
b) The side adjacent to angle B ?
c) The hypotenuse ?
d) Side b is ______ to angle A ?

Section 9-3: Using a Calculator to Find Trigonometric Ratios

If you know the acute angle measures of a right triangle, you can determine the sine, cosine, and tangent of each angle by using your calculator. For example, if you were given a right triangle with a $35°$ angle, you could determine the sine of that angle by doing the following on your calculator.

Scientific calculator: [sin] [35] [=]   OR   [35] [sin]]=]

Graphing calculator: [sin] [35] [ENTER]

The result would always be 0.5736 no matter what the size of the triangle.

Example 9-3.1: Given a right triangle that has an acute angle of $72°$.
　　　　　　　a)　　Find the value of the other acute angle.

　　　　　　　b)　　Find the ratio of the side adjacent to the $72°$ angle to the hypotenuse.

Solution:　　a)　Since the sum of the three angles in a triangle equals $180°$, the other acute angle is

$$180° - 90° - 72° = 18°$$

　　　　　　b)　The ratio that is in terms of the adjacent side and the hypotenuse is the cosine, so you can use the cos key on your calculator.

$$\cos 72° = \mathbf{0.309}$$

So in reference to the $72°$ angle, the ratio of the side adjacent to the hypotenuse is 0.309.　♦

A fact from geometry that can be useful says that in any triangle the greater side lies opposite the greater angle. Thus the hypotenuse is always larger than any of the other two sides.

Recall the three trigonometric ratios that we can use in relation to angle A in right triangle ABC:

$$\sin A = \frac{\text{opposite}}{\text{hypotenuse}}$$

$$\cos A = \frac{\text{adjacent}}{\text{hypotenuse}}$$

$$\tan A = \frac{\text{opposite}}{\text{adjacent}} \cdot$$

Activity 9.3

In the right triangle below, find

a)   the ratio of the side opposite $\angle$ A to the adjacent side if $\angle$ A $= 20°$.

b)   the ratio of the side opposite $\angle$ B to the hypotenuse if $\angle$ A $= 20°$.

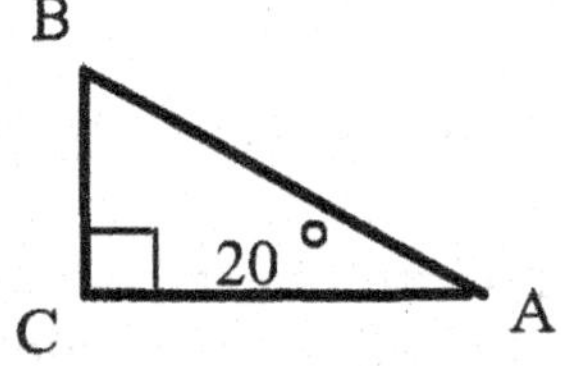

Example 9-3.2: Given a right triangle ABC, with A = 54° and b = 26 feet, find the length of the hypotenuse.

Solution: First let's sketch the right triangle.

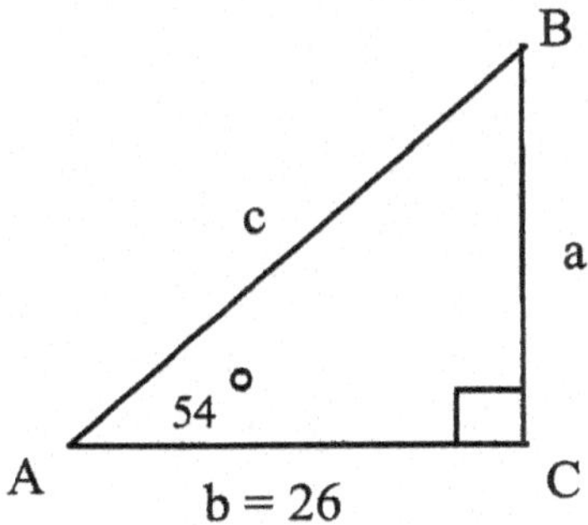

We are looking for the length of c, so using A as our reference angle; we know the **adjacent** side and are trying to find the **hypotenuse**. The ratio that is in terms of adjacent and hypotenuse is **cos**. So $\cos 54° = \dfrac{26}{c}$. Therefore, we need to solve for c to find the length of the hypotenuse.

$$\cos 54° = \frac{26}{c}$$

$$c(\cos 54°) = 26 \qquad \text{multiply both sides by c}$$

$$c = \frac{26}{\cos 54°} = 44.23 \qquad \text{divide both sides by } \cos 54°$$

So the hypotenuse is **44.23 feet**. ♦

Activity 9.4

a) Given a right triangle ABC, with $\angle A = 43°$ and c = 15, find the length of side a.

b) Given a right triangle ABC, with $\angle A = 36°$ and a = 20, find the length of side b.

Homework Sections 9-1 - 9-3

1. Given that $\angle A = \angle A'$, find x. $\angle C$ and $\angle C'$ are both right
    angles.

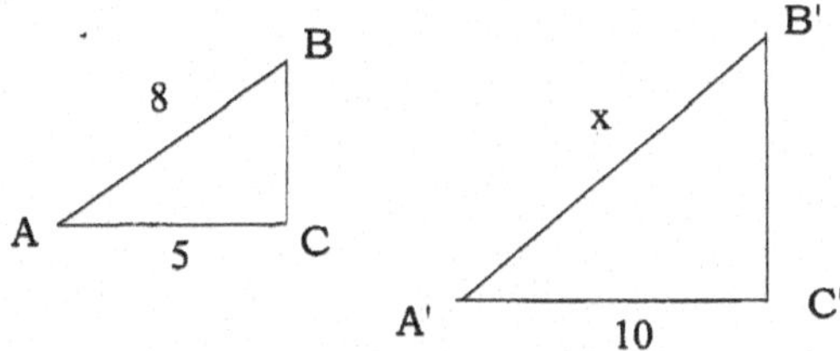

2. Given the triangle below, find the side adjacent to $\angle B$.

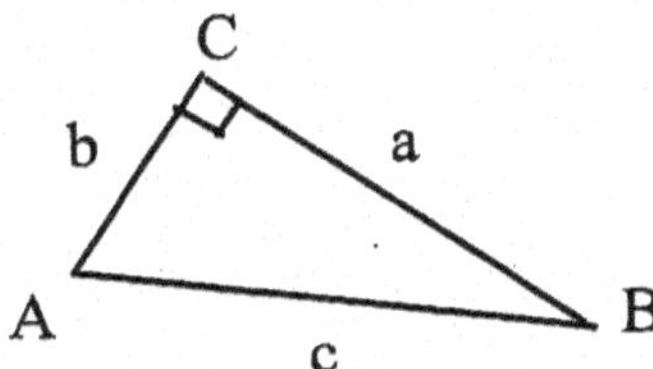

3. Given the triangle below, find the side opposite to $\angle A$.

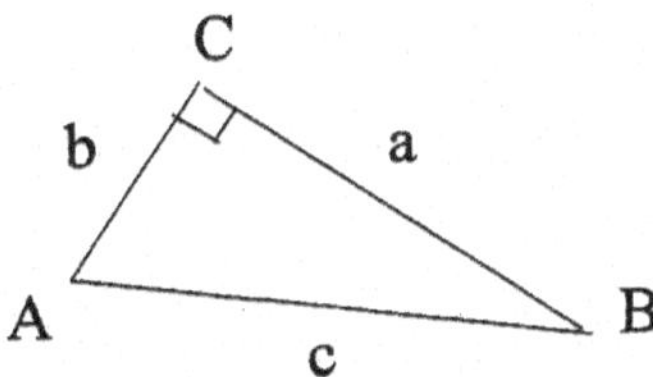

4. Given a right triangle that has an acute angle of $25°$.

    a)   Find the value of the other acute angle.

    b)   Find the ratio of the side opposite to the $25°$ angle to the hypotenuse.

5. Given a right triangle ABC, with $\angle B = 15°$ and c = 30, find the length of side a.

6. Given a right triangle ABC, with $\angle A = 50°$ and b = 30, find the length of side c.

<u>**Section 9-4: The Pythagorean Theorem**</u>

An important relation between the lengths of the sides of a right triangle is the **Pythagorean Theorem.** If you know the lengths for two sides of a right triangle, you can find the length of the remaining side using this relation.

> **Pythagorean Theorem:** For a right triangle with sides of length a, b, and c where c is the length of the hypotenuse,
> $$a^2 + b^2 = c^2.$$

Example 9-4.1: Use the Pythagorean Theorem to find the missing side of the following right triangle.

Solution:

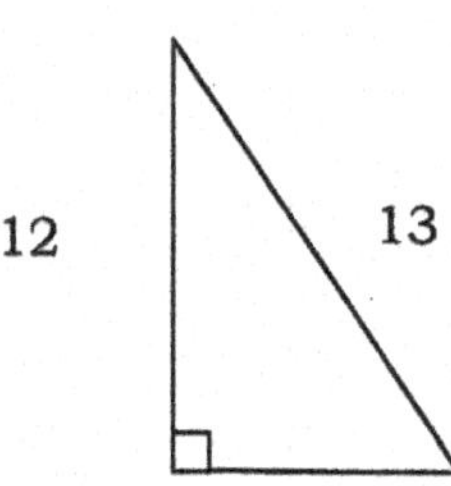

$x^2 + 12^2 = 13^2$ by the Pythagorean Theorem

$x^2 + 144 = 169$ square 12 and 13

$x^2 = 25$ subtract 144 from both sides

$x = 5$ take the square root of both sides

Note: When we take the square root of 25, we get +5 and -5, but because we are dealing with lengths, x ≠ -5.

Activity 9.5

Find the missing side for each of the following right triangles.

a)

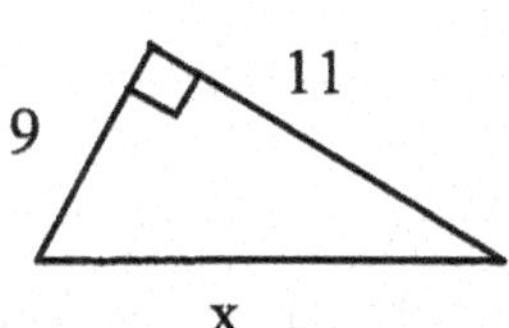

b)

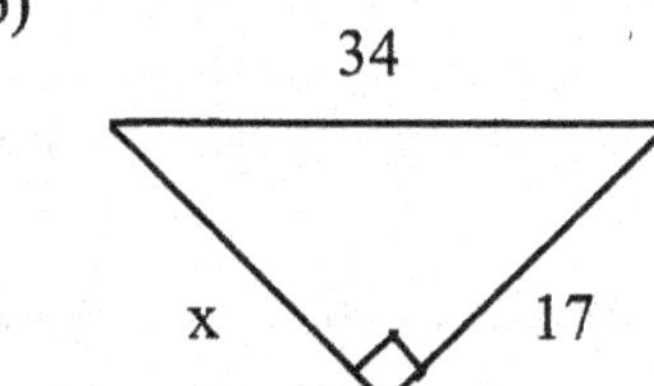

Section 9-5: Using Trigonometric Ratios to Determine Angles

In this section we will use the ratio of two sides of a right triangle to determine the angles of the triangle. Drawing and measuring triangles to determine the measure of angles is not precise and takes time. If we know the lengths of the sides of a right triangle, we can find the unknown angles by reversing the process of taking the sine, cosine, and tangent of an angle. We will need the following definitions.

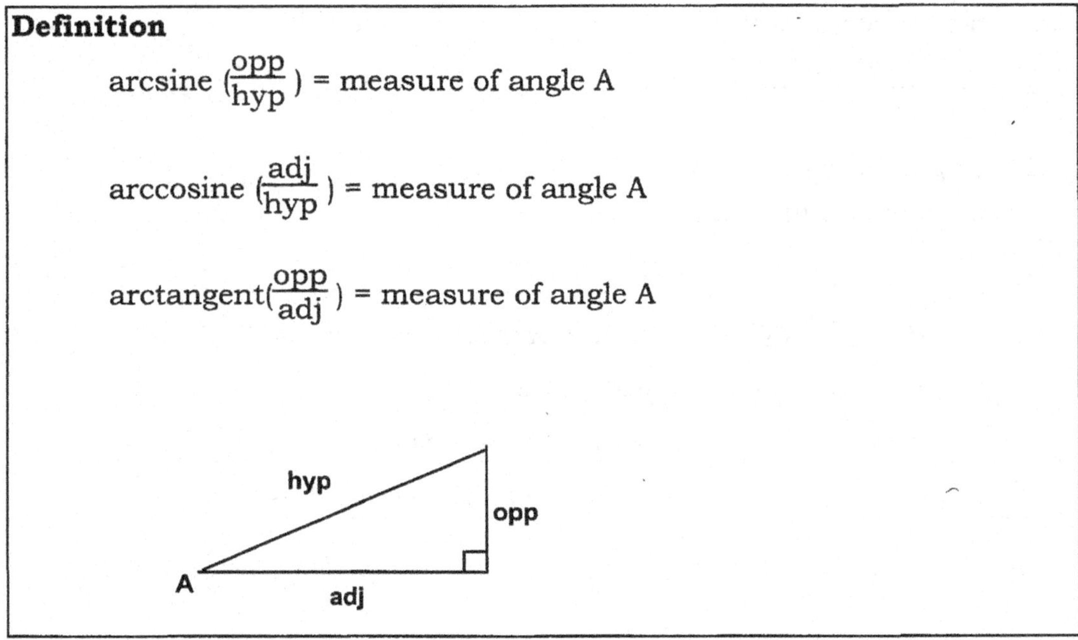

On your calculators, arcsine is denoted by $\sin^{-1}$, arccosine by $\cos^{-1}$, and arctangent by $\tan^{-1}$. This notation can be a little confusing at first because $\sin^{-1} A$ does not mean $\dfrac{1}{\sin A}$.

To understand the relationship between sine and arcsine look at the following.

$$\sin 65 = 0.906307787 \qquad \text{and} \qquad \sin^{-1} 0.906307787 = 65$$

$$\sin 12 = 0.2079116908 \qquad \text{and} \qquad \sin^{-1} 0.2079116908 = 12$$

You can see from these two examples that sine and arcsine undo each other the same way that x^2 and $\sqrt{}$ do. In other words, if you square a number and then take the square root of the answer you are back to the original number. This is useful because we know that if we take the sine of an angle, we get the ratio of the length of the opposite side to the length of the hypotenuse. By using the arcsine, if we know the ratio of the length of the opposite side to the length of the hypotenuse, we can get the angle.

Example 9-5.1:   Determine angle A in the following triangle.

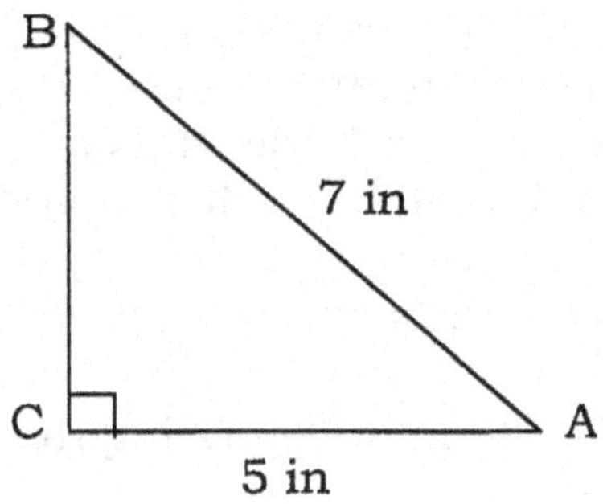

Solution: We are given the side adjacent to angle A and the hypotenuse.
So $\cos A = \dfrac{5}{7}$. From what we just saw, $\cos^{-1}\left(\dfrac{5}{7}\right) = A$. Evaluating
this with a calculator, we get $A = 44.4°$.   ♦

Suppose we knew the sine of some angle was 0.2588 but we did not know the angle.
To find the angle on the calculator we can use the SIN^{-1} key. You would enter the
following on your calculator:

Graphing calculator: $\boxed{\textbf{2nd}}$ $\boxed{\textbf{SIN}}$ $\boxed{.}$ $\boxed{\textbf{2 5 8 8}}$ $\boxed{\textbf{ENTER}}$

We get $A = 14.999°$.

Activity 9.6

a)   i)   Determine the measure of $\angle B$ in the following triangle.

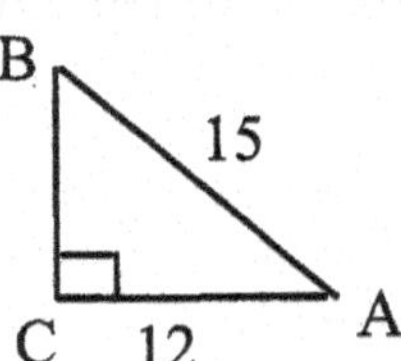

   ii)   Determine the length of a.

b)   If $\sin A = \dfrac{2}{3}$, find A.

Section 9-6: Degrees-Minutes-Seconds (DMS)

Rather than use decimal degrees (DD) to measure an angle, degrees may be divided into minutes, and minutes may be divided into seconds. One degree is equal to 60 minutes ($1° = 60'$), and 1 minute is equal to 60 seconds ($1' = 60''$). Sometimes it is necessary to convert from DMS notation to DD notation or vice versa.

Example 9-6.1: Change $67.3°$ to degrees, minutes, and seconds.

Solution: We need to convert $0.3°$ to minutes. Using the conversion $1° = 60'$ we can write

$$0.3 \text{ deg } \times \frac{60 \text{ min}}{1 \text{ deg}} = 0.3(60) \text{ min} = 18 \text{ min or } 18'.$$

So, $\mathbf{67.3°} = \boxed{\mathbf{67°18'}}$ ♦

Example 9-6.2: Change $27.34°$ to degrees, minutes, and seconds.

Solution: We first convert $0.34°$ to minutes.

$$0.34 \text{ deg } \times \frac{60 \text{ min}}{1 \text{ deg}} = 20.4 \text{ min}$$

Now we have to convert 0.4 min to seconds

$$0.4 \text{ min } \times \frac{60 \text{ sec}}{1 \text{ min}} = 24 \text{ sec}$$

So, $\mathbf{27.34°} = \boxed{\mathbf{27°20'24''}}$ ♦

Example 9-6.3: Change $9°43'20''$ to degrees in decimal form (round to the nearest thousandth of a degree).

Solution: First let's convert 20 seconds to minutes.

$$20 \text{ sec} \times \frac{1 \text{ min}}{60 \text{ sec}} = \frac{1}{3} \text{ min}$$

Therefore, $43'20'' = 43\frac{1}{3}$ min which now has to be converted to degrees.

$$\frac{130}{3} \text{ min } \times \frac{1 \text{ deg}}{60 \text{ min}} = \frac{130}{180} \text{ deg} = 0.722 \text{ degrees}$$

So, $\mathbf{9°43'20''} = \boxed{\mathbf{9.722°}}$ ♦

Activity 9.7

a)   Change $45.76°$ to degrees, minutes, and seconds.

b)   Change $12°25'30''$ to degrees in decimal form.

Homework Sections 9-4 - 9-6

1.   Find the missing side for each of the following right triangles.

a)   b)

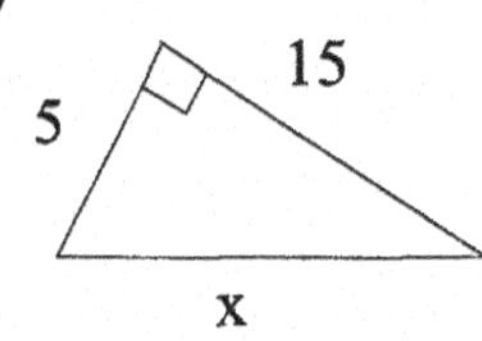
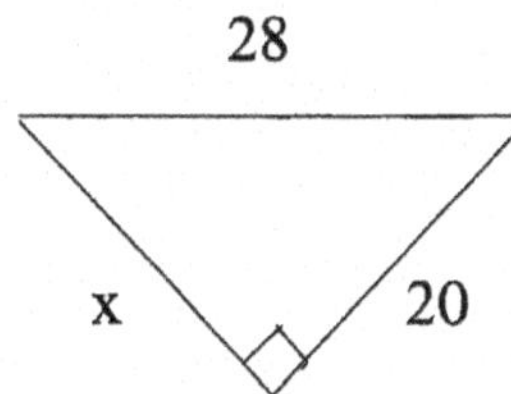

2.   a)

      i)   Determine the measure of $\angle B$ in the following triangle.

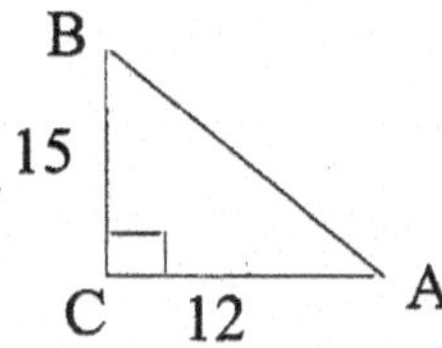

      ii)   Determine the length of c.

    b)   If $\sin A = \dfrac{5}{7}$, find A.

3.   Change 30.14° to degrees, minutes, and seconds.

4.   Change 24°10′30″ to degrees in decimal form.

<u>**Section 9-7: Applied Problems**</u>

Now that we have an understanding of the trigonometric ratios, let's look at how these ratios are used to solve applied problems.

Example 9-7.1:   In the right triangle ABC, $\angle A = 35°$ and c = 20 centimeters. Find a, b, and B.

Solution: **The first step in any applied problem is to make a diagram of the situation.** In this case, we get the following figure:

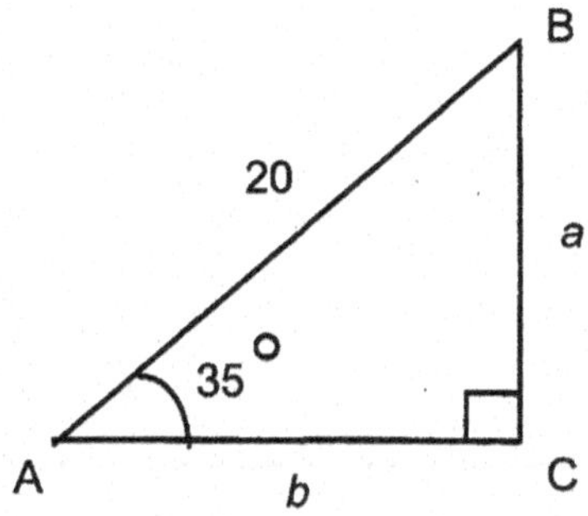

To find B, we use the fact that the sum of the two acute angles in any right triangle is $90°$.

$$B = 90° - A$$
$$= 90° - 35°$$
$$B = 55°$$

To find a, we can use the relationship $\sin A = \dfrac{a}{c}$.

$$\sin 35° = \frac{a}{20}$$
$$a = 20\sin 35° \qquad \text{multiply both sides by 20}$$
$$= 11.5$$

To find b, you could use the Pythagorean Theorem or you could use the relationship $\cos A = \dfrac{b}{c}$. Using this we get

$$\cos 35° = \frac{b}{20}$$

Solving for b we get

$$b = 20\cos 35° = 16.4$$

So a = 11.5, b = 16.4. and B = $55°$.   ♦

It is a good idea to check your answer using the Pythagorean Theorem, $a^2 + b^2 = c^2$. In the last example we came up with a = 11.5, b = 16.4, and c = 20. If we are correct $11.5^2 + 16.4^2$ should be approximately 20^2. In this case $11.5^2 + 16.4^2 = 401.2$, which is approximately $20^2 = 400$.

Activity 9.8

In the right triangle ABC, $\angle C = 90°$, $\angle A = 16°$ and a = 42 feet. Find b, c, and B.

Example 9-7.2:   A road rises 280 feet in elevation in a mile as measured on the road itself. What is the slope angle of the road in degrees?

Solution: First we have to draw a diagram.

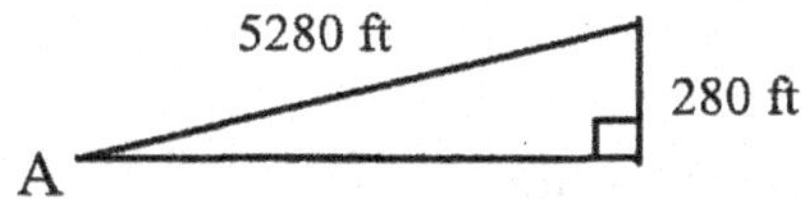

(Recall that 1 mile = 5280 ft)

We are looking for $\angle A$ in degrees. Since we know the length of the side opposite $\angle A$ and the hypotenuse, we have the following relationship:

$$\sin A = \frac{280}{5280}.$$

Since we know the ratio and we want to find the angle we need to use the arcsine. Recall that the arcsine is the $\sin^{-1}$ key on your calculator.

$$A = \sin^{-1}\left(\frac{280}{5280}\right)$$

$$A = 3.04°$$

Therefore, the slope angle of the road is $3.04°$.   ◆

Activity 9.9

A fifteen foot ladder leans against the top edge of a wall and makes an angle of 60°
with the ground.

a)   Find the distance between the foot of the ladder and the building.
        (Make a sketch first.)

b)   Calculate the height of the wall in at least two different ways.

For many application problems, we will need to understand the following definitions.

Definitions:
                If an object is located above the horizontal, then the angle
                between the horizontal and the line of sight XY is called the
                angle of elevation. (e in the triangle below)

                If the object is below the horizontal, then the angle between the
                horizontal and the line of sight XY is called the **angle of
                depression.** (d in the triangle below)

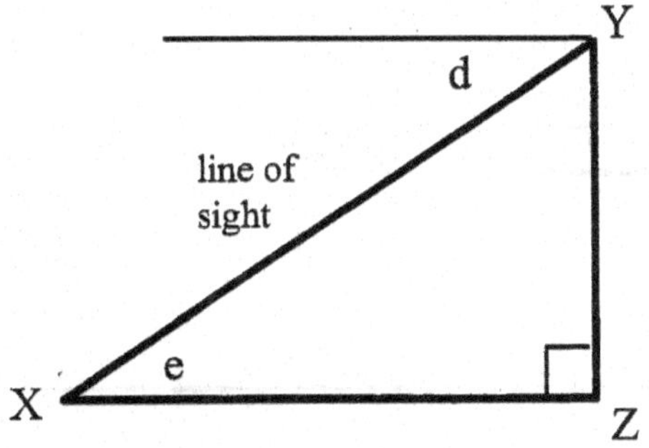

: A man climbs 800 feet up the side of a hill and finds that the angle of depression to his starting point is $50°$. How high is the hill?

Solution: First we need to draw a diagram.

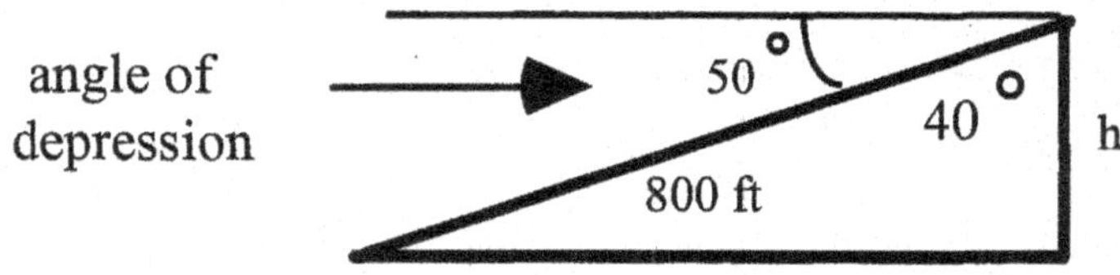

From the diagram we can write

$$\cos 40° = \frac{h}{800}$$

$$h = 800\cos 40° = 612.8 \text{ ft}$$

Therefore, the hill is 612.8 ft high.

♦

Activity 9.10

If a 43 foot flagpole casts a shadow 75 feet long, to the nearest 10 minutes what is the angle of elevation of the sun from the tip of the shadow?

The diagram for the problem is given below, finish the problem.

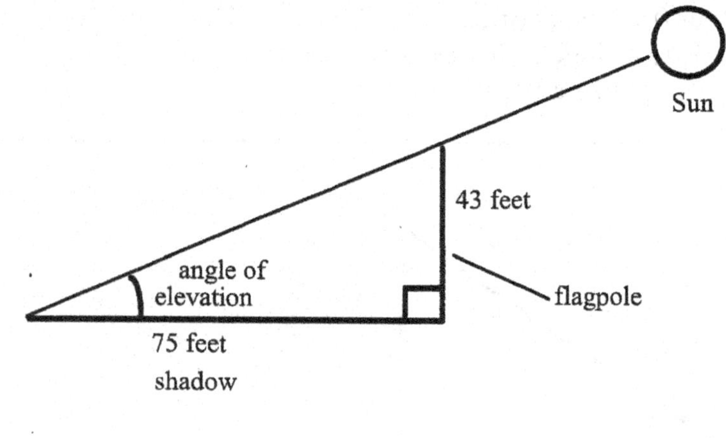

Water is often an obstacle to surveyors when measuring distances between two points. Suppose we had to measure the distance between points A and B in the figure below. A baseline AC perpendicular to AB is determined and then $\angle C$ is measured ($\angle C$ can also be referred to as $\angle ACB$). Then right triangle trigonometry can be used to determine the length of AB. (**Source:** G Rockswold, *Algebra and Trigonometry with Modeling and Visualization*)

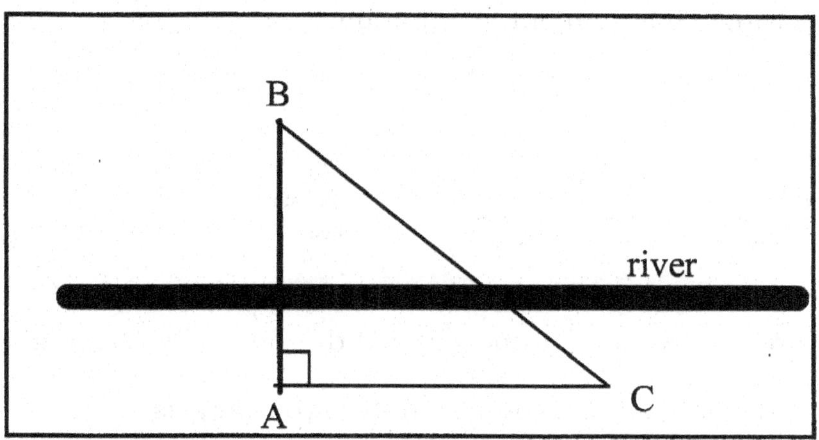

Activity 9.11

In the figure above, suppose the length of AC is 114.64 feet and $\angle C$ is 52.1°. Estimate the distance between points A and B.

Activity 9.12

Suppose you were driving along a straight level road toward a mountain range. While looking at one of the mountains ahead the passenger notices that the angle of elevation is $30°$. After driving a mile further, she measures the angle of elevation now to be 35°. How high is the mountain above the road?

Let's start off by drawing a diagram of the situation.

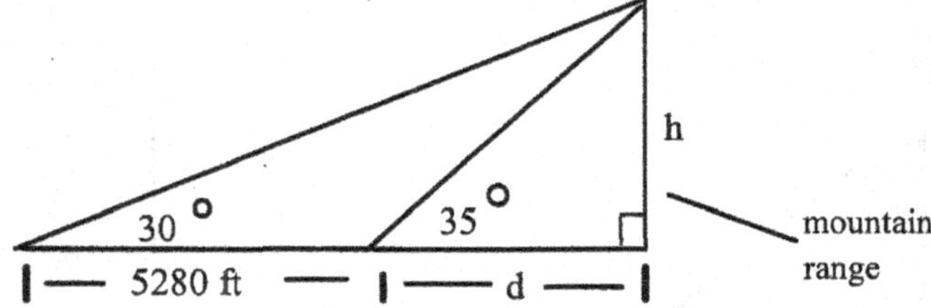

Notice that in this problem we do not know any of the sides of the triangles.

a)   Write down any relationships you see in the two triangles.

b)   Are there any common unknowns in any of your equations?

c)   Equate the two values for the common variable.

d)   Solve the resulting equation.

Unit 9 Review Exercises

1. When you find sin 23.8° using your calculator, you get .4035452964. To determine the accuracy to which the answer should be recorded, use the following rules.

Angles to the Nearest Degree	Significant Digits for Trigonometric Functions
One degree	2
Tenth of a degree	3
Hundredth of a degree	4
Thousandths of a degree	5
Ten-thousandths of a degree	6

Since 23.8° is measured to the nearest tenth of a degree, sin 23.8° should be reported with three significant digits, 0.404.

Compute the following values reporting answers to the correct number of significant digits.

a)   tan 47.98°   b)   cos 87°   c)   sin 7.9°

2. Find h in the following diagram.

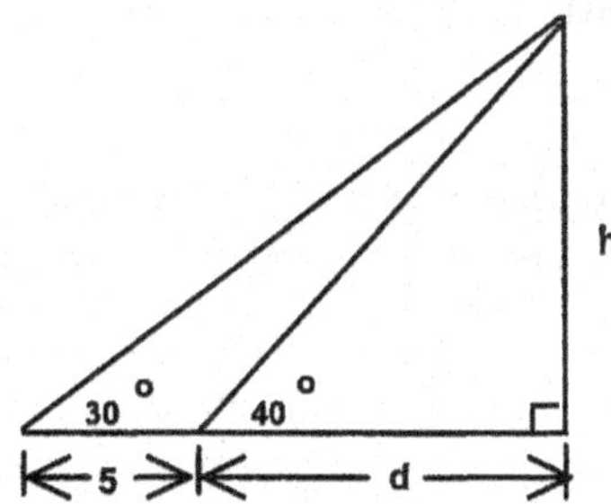

Solution: Using right triangle trigonometry, we have the following ratios:

$$\tan 40° = \frac{h}{d} \quad \text{and} \quad \tan 30° = \frac{h}{5 + d}.$$

Solve both of these equations for h and set the results equal to each other. Find d and use this result to find h.
Finish the problem.

3.	Find x in the following triangle. $\angle A = 25°$ and the hypotenuse is 10 units.

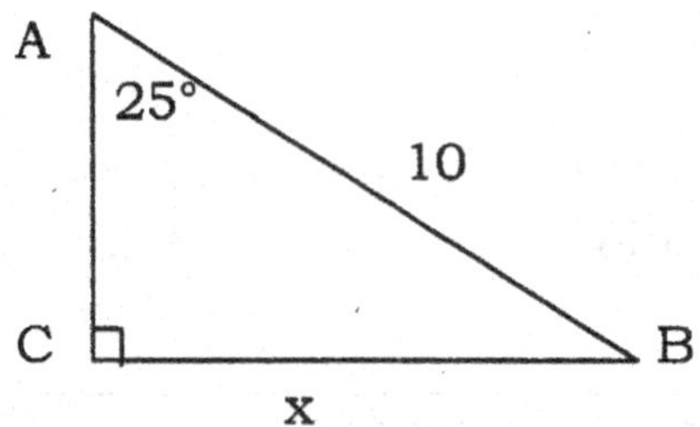

4.	Triangle ABC has a right angle at C. If $A = 42°$ and c = 12 feet, find b. Remember, start by drawing a diagram of the situation.

5.	Triangle ABC has a right angle at C. If a = 16 cm and b = 29 cm, find A.

6.	Triangle ABC has a right angle at C. If $B = 6°$ and a = 5.7 ft, find the missing parts of the triangle.

7.	A ramp for the disabled has to be built to the doorway of a house. The door is 3.0 feet above the ground and the ramp will make an angle of 5.0° with the ground. How long will the ramp be?

8.	Find the measure of $\angle A$.

a)	sin A = 0.8572					b)	cos A = 0.1392

9.	Suppose the maximum safe angle that a ladder can make with the ground is 65°. What is the shortest ladder you could use to reach a height of 35 feet on your house?

10.	Shown below are the rough plans for stairs going from the basement of your home to the outside. You have to build these stairs so compute the overall length of the stairs.

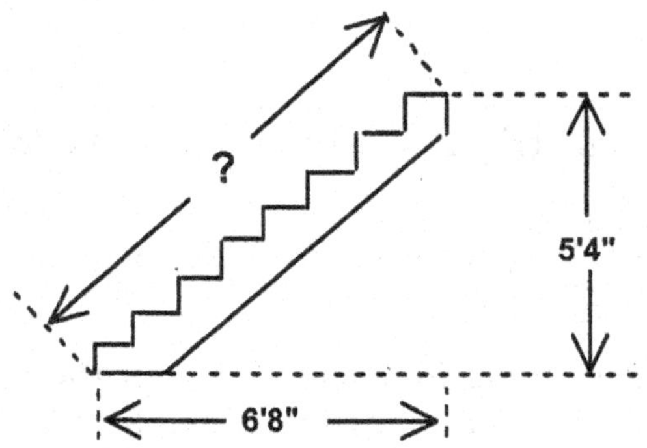

11.   An observer in a lighthouse 90 feet above the water measures the angle   of
      depression to a distant ship to be 1°. How many miles is the ship from the
      base of the lighthouse? (1 mile = 5280 feet)

12.   From a point on level ground, the angle of elevation to the top of a mountain is
      17°. Moving 200 meters closer the angle of elevation is now 24°. How high is
      the mountain?

13.   The Great Pyramid in Egypt has a square base 230 meters on each side. The
      faces of the pyramid make an angle of 51°50' with the horizontal.
      a)   What is the height of the pyramid?

      b)   What is the distance someone would have to climb to reach the
              top walking up a face?

      c)   Find the area of one of the faces in acres. (1 acre = 43, 560 sq ft)

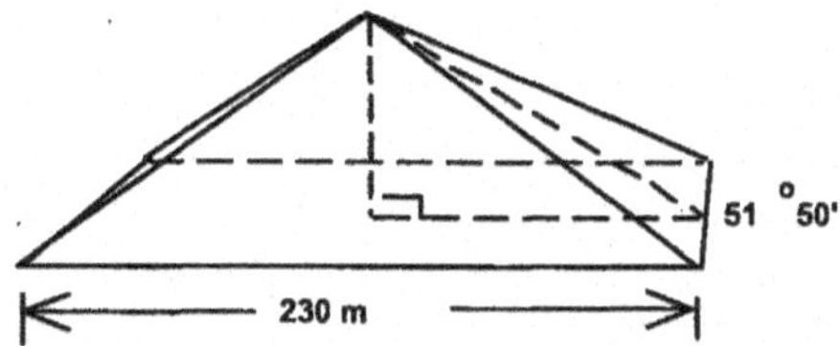

14.   Change the following to degrees, minutes, and seconds.

          i) 56.5°   ii)   14.46°

15.   Change the following to degrees in decimal form.

          i) 21°45'   ii)   19°40'10"

16.   Solve the right triangle which has the following given parts.

          A = 72.6°, c = 20

          <u>Sides</u>   <u>Angles</u>

          a = ____   A = 72.6

          b = ____   B = ____

          c = 20   C = 90

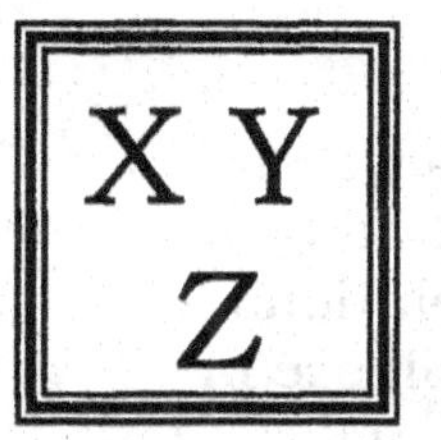

UNIT 10
Systems of Linear Equations

Section 10-1: Introduction

Many times when solving a problem, we will be required to deal with several quantities that are related to each other. Often the solution can be found by solving a **system of equations**. When you have to find all common solutions to two or more equations, the set of equations is called a **system**. You will see in subsequent sections how to solve a system of linear equations, but we will essentially be finding where two lines intersect.

Section 10-2: Solving a System of Linear Equations Using a Graph

To illustrate a system of linear equations, let's look at the following example.

The Math Club made $179 from a car wash. The members voted to use some money for a trip to Boston and some money to buy furniture for the Math Center. Let T = amount of money for the trip and F = amount of money for the furniture, then   T + F = 179.

Remember that every solution to this equation can be represented with an ordered pair. One solution is (50, 129), which means that $50 goes toward the trip and $129 goes for furniture.

Suppose further that the club decides that three times as much money should go toward the trip. This means that

$$T = 3F \text{ or } T - 3F = 0.$$

We want to know how much should go to each, the trip and the furniture. To solve this, we need to find an ordered pair that works in both of the equations.

$$T + F = 179$$
$$T - 3F = 0$$

Each of these equations can be represented with a line. The lines are shown in the graph below and the ordered pair we want is the intersection of the two lines. It appears that the solution is near (135, 45). So it looks like $135 will go toward the trip and $45 will be used for furniture.

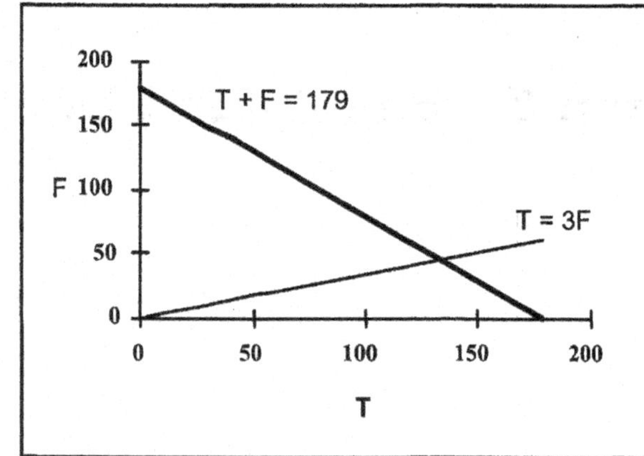

Checking our solution, we can see that $135 = 3(45)$ but $135 + 45 \neq 179$. So (135, 45) is only an approximation to our solution. This will usually be the case when you use a graph to solve a system of linear equations. To find exact solutions, we will use a few different algebraic methods.

When solving a system of linear equations there are three situations that can occur:

a) Intersecting lines:

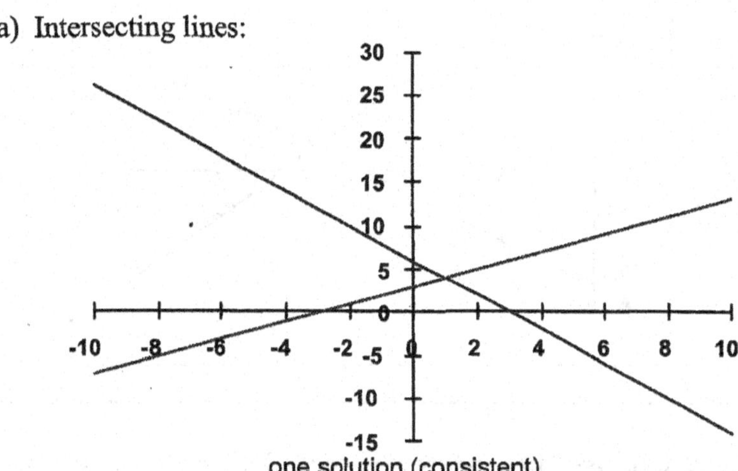

one solution (consistent)

b) Parallel Lines:

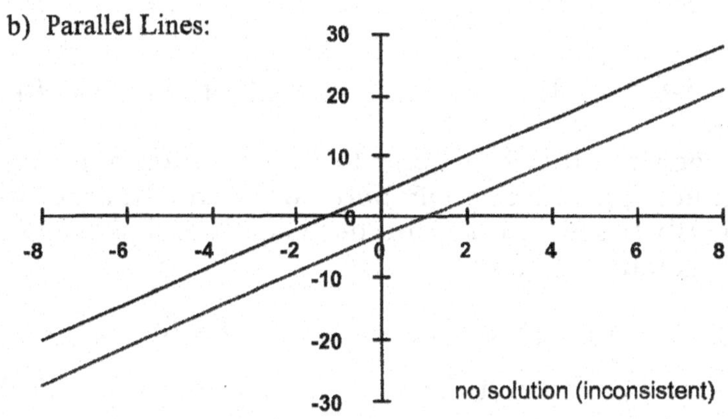

no solution (inconsistent)

c) Coinciding Lines:

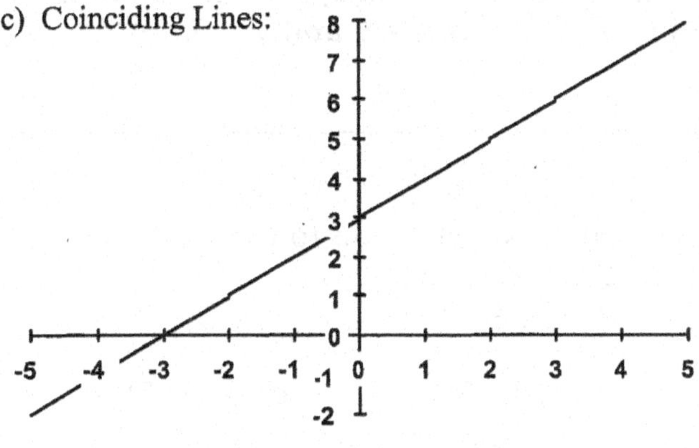

many solutions (dependent)

Approximate the solution to the system of equations whose graphs are shown below.

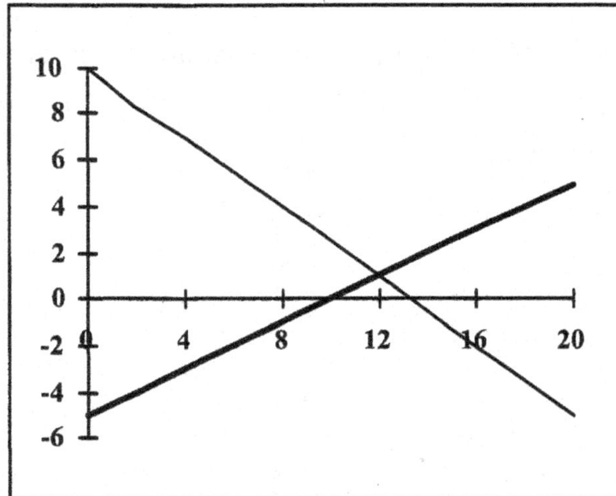

Example 10-2.1: Choose the ordered pair which is the solution to the given system of linear equations.

$$x + y = 16$$
$$2x - y = 8$$

a) $(5, 11)$ b) $(6, 4)$ c) $(8, 8)$ d) $(6, 10)$

Solution: In order for an ordered pair to be a solution to a system of linear equations, it must satisfy both equations. In other words, if you substitute x and y into both equations, the two resulting statements will be true.

a) $5+11=16$ b) $6+4\neq16$ c) $8+8=16$ d) $6+10=16$

$2(5)-11\neq8$ $2(6)-4=8$ $2(8)-8=8$ $2(6)-10\neq8$

Therefore, the solution to the system of equations is the ordered pair **(8, 8)** which means x = 8 and y = 8.

♦

Activity 10.2

Choose the ordered pair, which is a solution to the given system of equations.

$$2x + 5y = -9$$
$$x + 3y = -2$$

a) $(-41, 13)$ b) $(13, -5)$ c) $(-17, 5)$ d) $(37, -13)$

We previously saw that graphing usually gives us an approximate solution to a system of equations. One method that gives an exact solution is **substitution**.

Substitution Method

1. Solve one equation for one of the variables.

2. Substitute the result in the other equation.

3. Now you have one equation with one variable, solve the equation for that variable.

4. Substitute that value in the first equation and solve that equation for the second variable.

5. Check your solutions.

6. State the solutions.

Example 10-3.1: Use substitution to solve the following system of equations.

$$x - y = 10$$
$$2x - 3y = 4$$

Solution: Step 1.   Solve one equation for one variable.

$$x - y = 10 \qquad \text{solving this equation for x we get}$$
$$x = 10 + y$$

Step 2.   Substitute $10 + y$ for x in the other equation.

$$2x - 3y = 4$$
$$2(10 + y) - 3y = 4$$

Step 3.   Solve for y.

$$2(10 + y) - 3y = 4$$
$$20 + 2y - 3y = 4$$
$$-y = -16$$
$$\mathbf{y = 16}$$

Step 4.   Substitute 16 for y in the other equation and solve for x.

$$x - 16 = 10$$
$$\mathbf{x = 26}$$

Step 5.   Check the solution x = 26 and y = 16.

$$26 - 16 = 10$$
$$2(26) - 3(16) = 4$$

Step 6.   The solution to the system of equations is **x = 26** and **y = 16**. When graphed these 2 equations are intersecting lines that meet at the point (26, 16).

◆

When solving a system of linear equations using substitution you want to make sure in step 1 that you choose the equation and variable that will give the easiest expression for substitution. For example, in the system above if you solved equation 2 for x you would have obtained $x = \dfrac{4 + 3y}{2}$ which would be more difficult to work with.

Example 10-3.2: Use substitution to solve the following system of equations.

$$2x + 4y = 29$$
$$x - y = -2$$

Solution: Step 1.   Solve one equation for one variable.

x - y = -2   solving this equation for x we get
x = -2 + y

Step 2.   Substitute -2 + y for x in the other equation.

$$2x + 4y = 29$$
$$2(-2 + y) + 4y = 29$$

Step 3.   Solve for y.

$$2(-2 + y) + 4y = 29$$
$$-4 + 2y + 4y = 29$$
$$6y = 33$$
$$\mathbf{y = 5.5}$$

Step 4.   Substitute 5.5 for y in the other equation and solve for x.

$$x - 5.5 = -2$$
$$\mathbf{x = 3.5}$$

The solution to the system of equations is **x = 3.5** and **y = 5.5**. Now check to make sure (3.5, 5.5) satisfies both equations.

◆

Solve the following system of equations by substitution.

$$3x - 2y = 0$$
$$x - 3y = -7$$

Homework Sections 10-2 and 10-3

1. Approximate the solution to the system of equations whose graphs are shown below.

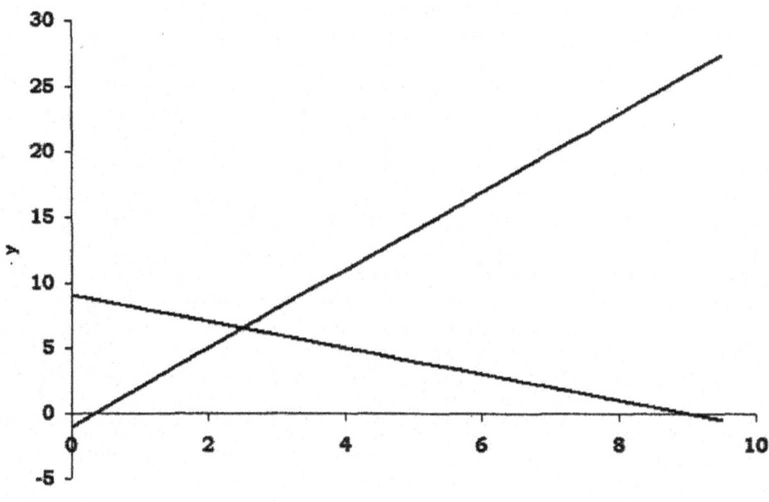

2. Which of the following is a solution to the system of equations?

$$4x - 3y = 16$$
$$x - y = 5$$

a) (2, -3)   b) (1, -4)   c) (4, 0)   d) (7, 4)

3. Solve the following systems of equations by substitution.

a) $x + y = 5$
      $x - y = 1$

b) $3x + 2y = 3$
      $x + 3y = 8$

c) $2x - y = 0$
      $3x + 2y = 14$

Another way to find a solution of a system of linear equations algebraically is to use the properties of equality that were used earlier to solve equations.

If equals are added or subtracted from equals, then the results are equal.

Example 10-4.1: Suppose you had the following system of equations:

$$2x + 3y = 12$$
$$9x - 3y = 10$$

Let's **add** the left sides and the right sides, leaving us with

$$2x + 3y = 12$$
$$\underline{9x - 3y = 10}$$
$$\mathbf{11x + 0 = 22}$$

(This worked out nice because now we have one equation and one unknown which we can solve.)

Solving this for x, we get $x = 2$.

Substituting 2 for x in the first equation we get

$$2(2) + 3y = 12$$
$$4 + 3y = 12$$
$$3y = 8$$
$$y = \frac{8}{3}$$

Hence the solution to the system of equations is $x = 2$ and $y = \dfrac{8}{3}$.

Remember the solution, $\left(\mathbf{2}, \dfrac{\mathbf{8}}{\mathbf{3}}\right)$, is the point where the graphs of the two lines intersect. ◆

Suppose you had the following system of equations and added them the way we did in the last example, notice that we are left with one equation and two unknowns. This equation has an infinite number of solutions.

$$3x - 2y = 9$$
$$\underline{5x + 7y = 12}$$
$$\mathbf{8x + 5y = 21}$$

When we add the equations we want to end up with just one equation and one unknown. The following steps summarize how to solve a system like the one above using the addition method.

Addition Method

1. Put both equations in the form Ax + By = C.

2. Decide which variable to eliminate. Then multiply one or both equations by the number such that the coefficients of the variable to be eliminated will be the opposite of each other.

3. Add the two equations to obtain a new equation in one variable.

4. Solve the new equation for the one variable.

5. Substitute this value into one of the original equations and solve for the second variable.

6. Check your solutions.

7. State the solutions.

Example 10-4.2: Solve the following linear system using the addition method.

$$3x - y = 7$$
$$2x + 3y = 12$$

Solution: Step 1. The equations are already in the form Ax + By = C.

Step 2. If we multiply the first equation by 3, the coefficients of y will be opposites.

$$9x - 3y = 21$$
$$2x + 3y = 12$$

Step 3. Add the two equations.

$$11x = 33$$

Step 4. Solving for x we get:

$$x = 3$$

Step 5. Substitute this value into the first equation and solve for y.

$$3(3) - y = 7$$
$$y = 2$$

Step 6.   Check the solution x = 3 and y = 2.

$$3(3) - 2 = 7$$
$$2(3) + 3(2) = 12$$

Step 7.   The solution is **x = 3** and **y = 2**.   ♦

Sometimes, you may have to multiply both equations to get one of the variables to have opposite coefficients. This is illustrated in the next example.

Example 10-4.3: Solve the following system of equations.
$$2x + 4y = 7$$
$$-3x + 3y = 2$$

Solution: Step 1.   Both equations are in the form Ax + By = C.

Step 2.   Let's eliminate x by multiplying the first equation by 3 and the second by 2, this will give opposite coefficients for x.

$$6x + 12y = 21$$
$$-6x + 6y = 4$$

Step 3.   Add the resulting equations.

$$18y = 25$$

Step 4.   Solve for y.

$$y = \frac{25}{18}$$

Step 5.   Substitute this value into one of the original equations and solve for x.

$$2x + 4\left(\frac{25}{18}\right) = 7 \qquad \text{(Multiply both sides by 18.)}$$

$$36x + 100 = 126$$

$$36x = 26$$

$$x = \frac{26}{36} = \frac{13}{18}$$

Step 6.   Check the solution to make sure $(\frac{13}{18}, \frac{25}{18})$

satisfies both of the equations.

Step 7.   The solution is $x = \frac{13}{18}$ and $y = \frac{25}{18}$.   ◆

In the last example, rather than substitute $y = \frac{25}{18}$ into one of the original
equations, it may have been easier to go back to the original equations and
eliminate y. In the next activity you can try this.

Activity 10.4

In the following system of equations, eliminate y and solve for x. You should get
the same solution as the last example.

$$2x + 4y = 7$$
$$-3x + 3y = 2$$

When we solve a system of equations like

$$2x + 4y = 7$$
$$-3x + 3y = 2$$

we are looking for the point where the two lines intersect. If we multiply the first
equation by 3 and the second equation by -4 we get the resulting equations

$$6x + 12y = 21$$
$$12x - 12y = -8.$$

This system of equations is different than the original system, but the thing to
remember is, the graphs are the same. In other words, $2x + 4y = 7$ and
$6x + 12y = 21$ have the same graphs. This means the second system above has
to have the same solution as the first system since the graphs are the same;
they have to intersect in the same place.

We said at the beginning of the unit that three things can happen when you are solving a system of linear equations. You can have a unique solution (the lines intersect in a point), no solution (the lines are parallel), an infinite number of solutions (the lines are the same). So far, all the examples have resulted in a unique solution. How do we know when we have no solution or an infinite number of solutions? If you are solving a system of equations algebraically and the resulting equation is 0 = non-zero, the equations have no-solution, if the resulting equation is 0 = 0, the system has an infinite number of solutions.

Example 10-4.4:   Solve the following system of equations.
$$15x + 12y = 8$$
$$10x + 8y = 13$$

Solution: We will solve the system using the **addition method**.

$$15x + 12y = 8$$
$$10x + 8y = 13$$

If we eliminate the x, we can multiply equation 1 by 10 and multiply equation 2 by -15 giving us

$$150x + 120y = 80$$
$$-150x - 120y = -195$$

Adding the two equations, we get

$$0 + 0 = -115 \text{ or } 0 = -115$$

This tells us that there is no-solution to the system of equations. ♦

Example 10-4.5: Solve the following system of equations.
$$-6x - 2y = -12$$
$$3x + y = 6$$

Solution: We will solve the system using the **substitution method**. Solving for y in equation 2, we get

$$y = -3x + 6$$

Substituting y into equation 1 gives us

$$-6x - 2(-3x + 6) = -12$$

Simplifying, we get

$$-6x + 6x - 12 = -12$$

$$0 = 0$$

This tells us that the system of equations has an infinite number of solutions.

If you look closely at the two equations, you can see that equation 1 is just a multiple of equation 2. When 2 equations in a system are multiples, they will have infinitely many solutions.

Excel makes solving linear systems of equations using **determinants** very convenient. A **determinant** is a unique value that is associated with a square array of numbers called a **matrix**. Given a square array of numbers

$A = \begin{bmatrix} a & b \\ c & d \end{bmatrix}$, the determinant of A is denoted by

$$\det A \quad \text{or} \quad \begin{vmatrix} a & b \\ c & d \end{vmatrix} .$$

$$\det A = \begin{vmatrix} a & b \\ c & d \end{vmatrix} = ad - cb$$

Example 10-5.1: Evaluate the determinant $\begin{vmatrix} 1 & 3 \\ 5 & -2 \end{vmatrix}$

Solution: By the definition

$$\begin{vmatrix} 1 & 3 \\ 5 & -2 \end{vmatrix} = (1)(-2) - (5)(3)$$

$$= -2 - 15 = -17 \qquad \blacklozenge$$

Activity 10.6

Evaluate the following determinants.

a) $\begin{vmatrix} 2 & 6 \\ -1 & 5 \end{vmatrix}$

b) $\begin{vmatrix} -1 & 1 \\ 5 & 2 \end{vmatrix}$

Cramer's Rule is another technique used to solve linear systems of equations.

Cramer's Rule

The solution to the linear system of two equations and two variables:
$$ax + by = c$$
$$dx + ey = f$$
is given by

$$x = \frac{\begin{vmatrix} c & b \\ f & e \end{vmatrix}}{\begin{vmatrix} a & b \\ d & e \end{vmatrix}} \qquad\qquad y = \frac{\begin{vmatrix} a & c \\ d & f \end{vmatrix}}{\begin{vmatrix} a & b \\ d & e \end{vmatrix}}$$

Cramer's Rule may look hard to remember, but if you look carefully at it you can see a pattern. For both x and y the denominator is just the coefficients for x and y. In the numerator when you are finding x, you replace the x column with the constant column and the y column does not change. When you are finding y, the y column gets replaced with the constant column and the x column remains unchanged.

Example 10-5.2: Use Cramer's Rule to solve the following linear system.
$$3x - y = 7$$
$$2x + 3y = 12$$

Solution:

$$x = \frac{\begin{vmatrix} 7 & -1 \\ 12 & 3 \end{vmatrix}}{\begin{vmatrix} 3 & -1 \\ 2 & 3 \end{vmatrix}} = \frac{21-(-12)}{9-(-2)} = \frac{33}{11} = 3$$

$$y = \frac{\begin{vmatrix} 3 & 7 \\ 2 & 12 \end{vmatrix}}{11} = \frac{36-14}{11} = \frac{22}{11} = 2$$

The solution to the system of equations is **x = 3** and **y = 2**. ♦

We can use Excel to find the determinant of a square matrix. Let's use the steps below to evaluate $\begin{vmatrix} 3 & 7 \\ 2 & 12 \end{vmatrix}$ from the previous example.

1. Enter the given square matrix.
2. Highlight a cell of the worksheet (near the given matrix) where you wish the value of the determinant to appear.
3. Type: = **MDETERM(B2:C3)** (This will appear in the formula bar.)
4. Press **ENTER** to get the result.

	A	B	C	D	E	F
1						
2		3	7			
3		2	12		determinant =	=MDETERM(B2:C3)

The answer of 22 will appear if cell F3.

Activity 10.7

Solve the following system of equations using Cramer's rule and Excel.

$$2x - 7y = 1$$
$$-3x - y = 10$$

When using **Cramer's Rule** if $x = \dfrac{non - zero}{zero}$ the system has no-solution, and if $x = \dfrac{zero}{zero}$ the system has an infinite number of solutions.

Activity 10.8

Solve the following system of equations using Cramer's rule and Excel.
(Hint: remember to put the equation in the form Ax + By = C.)

$$y = 5 - 3x$$
$$9x + 3y = 15$$

Homework Sections 10-4 and 10-5

Solve the following systems of equations by both the addition method and Cramer's rule.

1. $6x - y = 5$
    $3x + 4y = 2$

2. $5x + 3y = 3$
    $2x + 3y = 4$

3. $3x + 4y = 6$
    $2x - 6y = 5$

4. $0.7x + 5.3y = 6.6$
    $5.2x + 2.2y = 1.7$

5. $2x - 7y = 9$
    $4x - 14y = 18$

6. $x - 5y = 0$
    $-3x + 15y = 9$

Supply and **demand** is perhaps the most basic law in economics. This law is usually illustrated by two curves: one showing the demand, which represents what you as a consumer will do, and one showing the supply, which represents what the company will do. Usually the two curves are shown as straight lines as shown in the graph below.

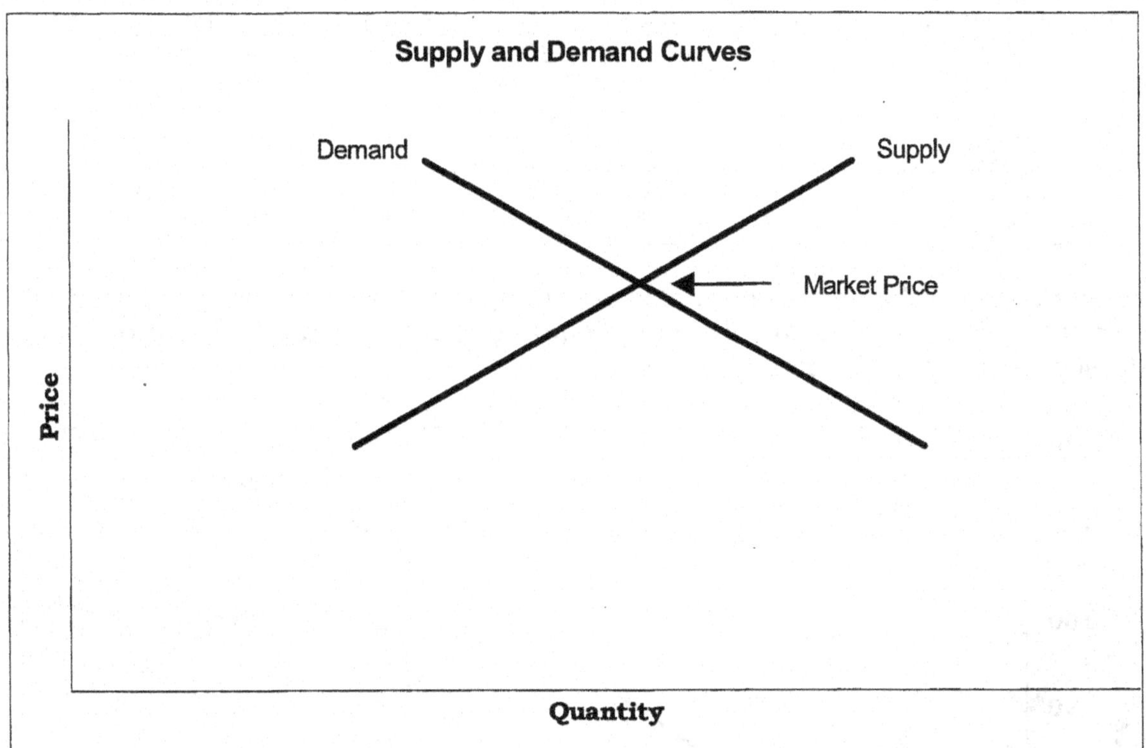

Demand refers to how much (quantity) of a product or service is desired by buyers. The quantity demanded is the amount of a product people are willing to buy at a certain price; the relationship between price and quantity demanded is known as the demand relationship. In general, the lower the price, the more likely that you will buy an item, which explains the negative slope in the diagram.

Supply represents how much the market can offer. The quantity supplied refers to the amount of a certain good producers are willing to supply when receiving a certain price. The correlation between price and how much of a good or service is supplied to the market is known as the supply relationship. The higher the price, the more likely the supplier will produce because it will make more money.

The point where the two curves intersect is called the **price equilibrium** or the **market price**. Equilibrium is a situation in which there are no inherent forces that produce change. Changes away from an equilibrium position will occur only as a result of outside events that disturb the status quo.

Consider the following table of sales of bottles water at a local convenience store: (modified from *Teaching Money Applications to Make Mathematics Meaningful* by Marquez and Westbrook).

Quantity of Bottles Demanded	Quantity of Bottles Supplied	Price per Bottle
100	220	130 cents
120	200	120
140	180	110
160	160	100
180	140	90
200	120	80

The equation for the demand curve is P = -0.5Q + 180 and the supply equation is P = 0.5Q + 20. Solving this system of equations will give us the point of intersection (160, 100) or (160, $1.00).

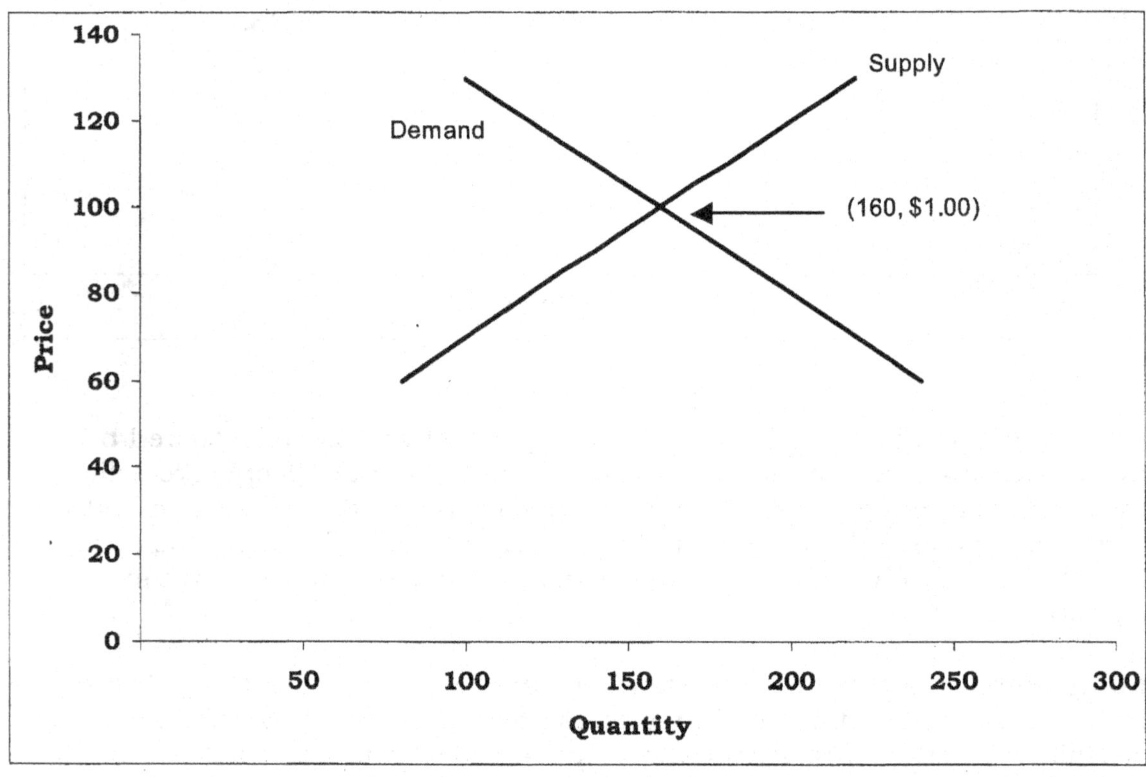

Suppose there was a period of hot weather causing an increase of 100 bottles in demand at every price level. The figure below shows a translation of the demand curve to the right.

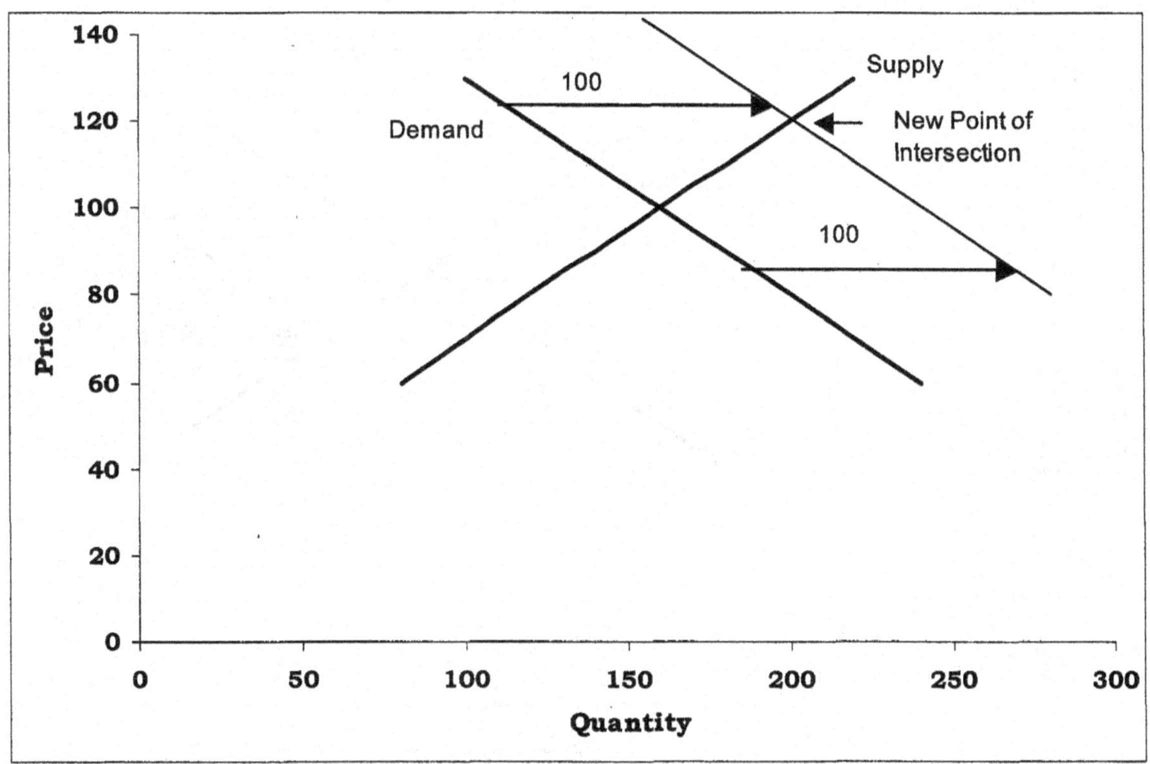

The new demand curve becomes P = -0.5(Q – 100) + 180 or simplifying P = -0.5Q + 230.

Activity 10.9

a) Using the above graph, estimate the new point of intersection.

b) Find the exact point of intersection algebraically by using the new demand curve given above and the supply equation given on the previous page.

Suppose there is a new store opening in the area which causes the supply to increase, thus moving the supply curve to the right. You can see in the figure below that moving the supply curve to the right results in a decrease in market price.

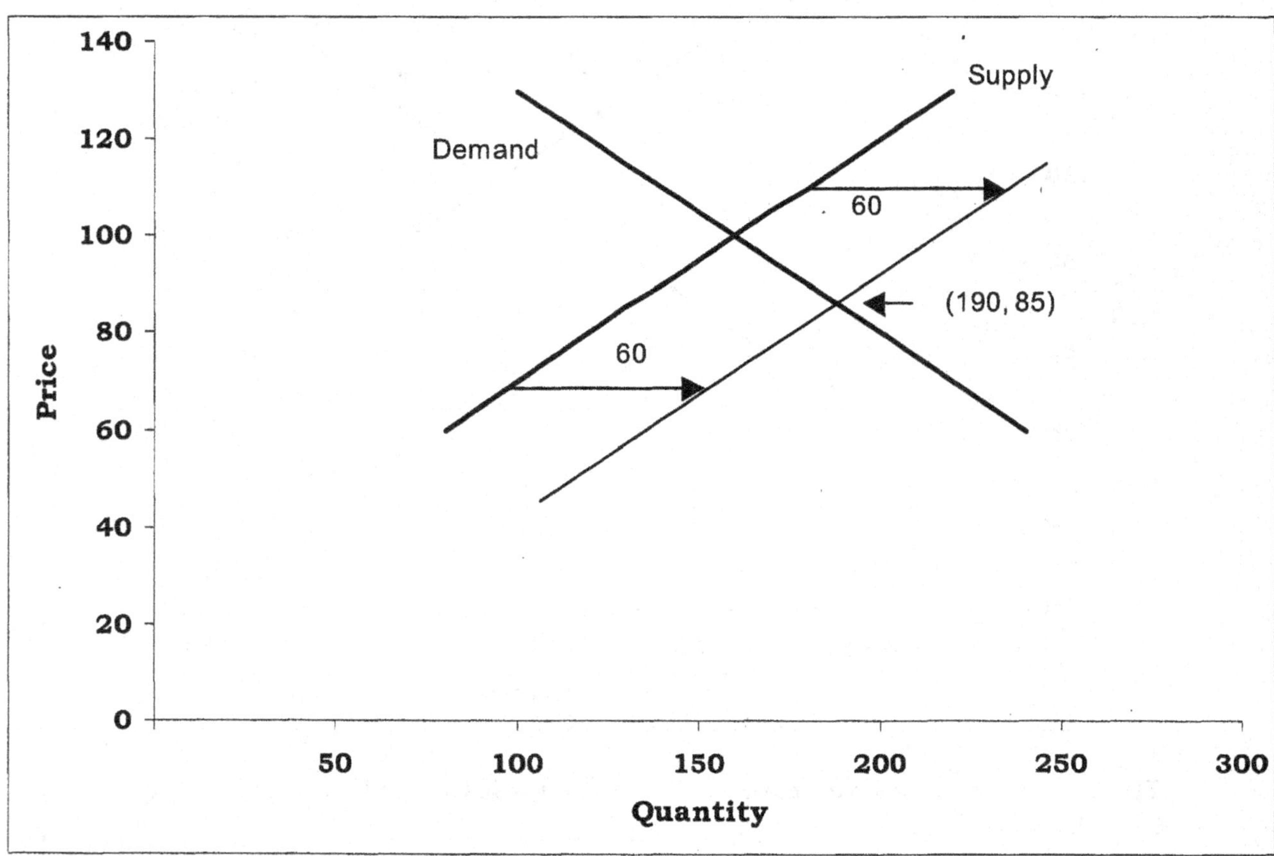

Activity 10.10

a)   If the supply curve translates 60 units to the right, find the new equation for the supply curve.

b)   Algebraically, verify that the new point of intersection given in the graph is correct.

The last few examples illustrate the **law of supply and demand**, which states that, in a free market, the forces of supply and demand generally push the price toward the level at which quantity supplied and quantity demanded are equal.

As an application of the law of supply and demand, consider the following example. Suppose the state legislature raises the gasoline tax by 28 cents per gallon. Gas station owners will then have to collect 28 additional cents in taxes on every gallon they pump. The gas station owners would like to shift the entire tax to buyers, but as you will see the market mechanism allows them to shift only part of the 28 cents.

In the figure below, the demand curve is the curve D. The supply curve before the new tax is the curve S_0 and before the new tax the equilibrium point is E_0, (50, \$2.67). With the new tax, at each quantity the gas station owners have to receive an additional 28 cents, which moves S_0 to S_1. Notice now the equilibrium point moves from E_0 to E_1. The price increases from \$2.67 to \$2.75, telling us that the price of gas only increased by 8 cents, not 28 cents. This means that the gas station owners would have to make up the additional 20 cents.

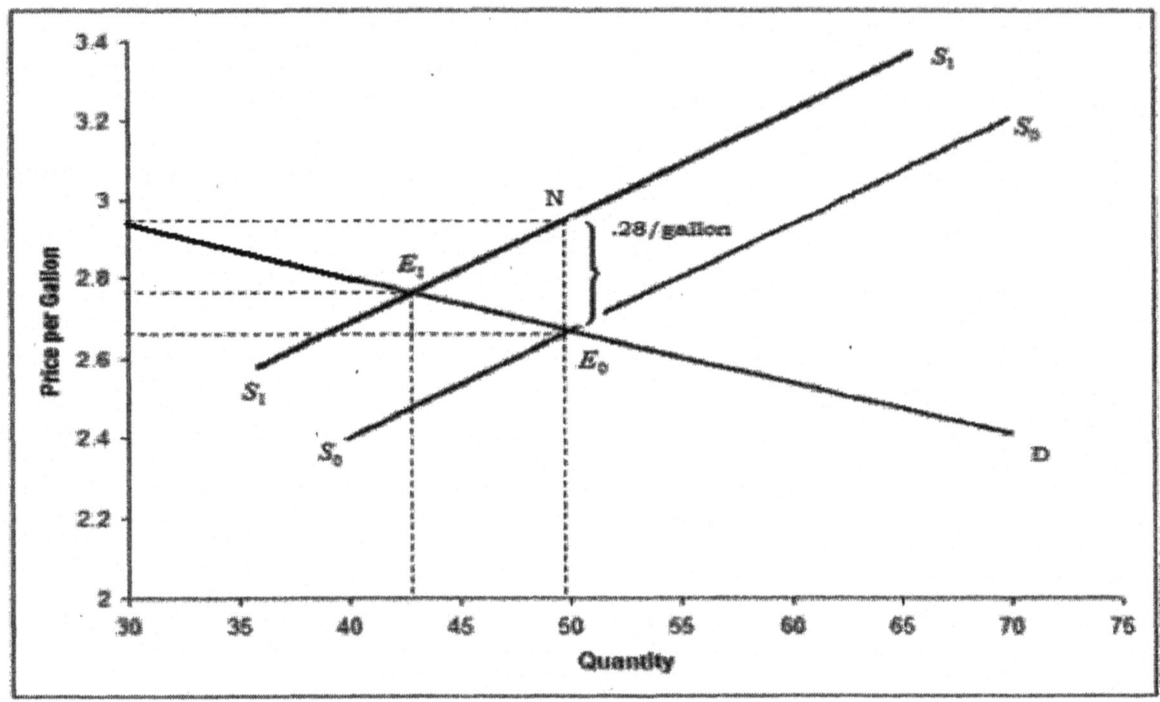

This example shows the cost of almost any commodity will usually be paid partly by the consumer, partly by the seller.

Activity 10.11

Make some observations about supply and supply-demand equilibrium.

For example, you could say that any change that shifts the supply curve outward to the right, and does not affect the demand curve, will lower the equilibrium price and raise the equilibrium quantity.

Unit 10 Review Exercises

1. Solve the following system of equations by the substitution method **or** by the addition method.

   a) $y = 3x$
       $5x - y = 7$

   b) $4x + 6y = 9$
       $x - 2y = 10$

   c) $3x - 5y = 8$
       $2x - 6y = 1$

   d) $y = 3x + 4$
       $3x + 5y = 20$

2. Explain what happens when you try to solve the following system of equations.

$$x + 3y = 4$$
$$4x + 12y = 12$$

3. Evaluate the following determinants.

   a) $\begin{vmatrix} 1 & 3 \\ 4 & -2 \end{vmatrix}$

   b) $\begin{vmatrix} 1 & -4 \\ 3 & 2 \end{vmatrix}$

4. Solve the following system of equations using

   a) the addition method   b) Cramer's rule.

$$4r - 6t = 9$$
$$6r + 2t = 11$$

5. Extend Cramer's rule to solve the following system of equations. (Use Excel to evaluate the determinants).

$$3x - 8y + z = 9$$
$$x + 2y - 3z = 0$$
$$5x + 5y + 2z = -4$$

6. The demand and supply curves for Keene State shirts are given by the following equations:

$$Q = 3,000 - 20P \qquad\qquad Q = 750 + 40P$$

where P is measured in dollars and Q is the number of shirts sold per year.

a) Find the equilibrium price and quantity.

b) Graph the supply-demand curves.

7. Using the graph below, approximate the solution to the following system of equations:

$$3x + 7y = 9$$
$$4x - 6y = -2.$$

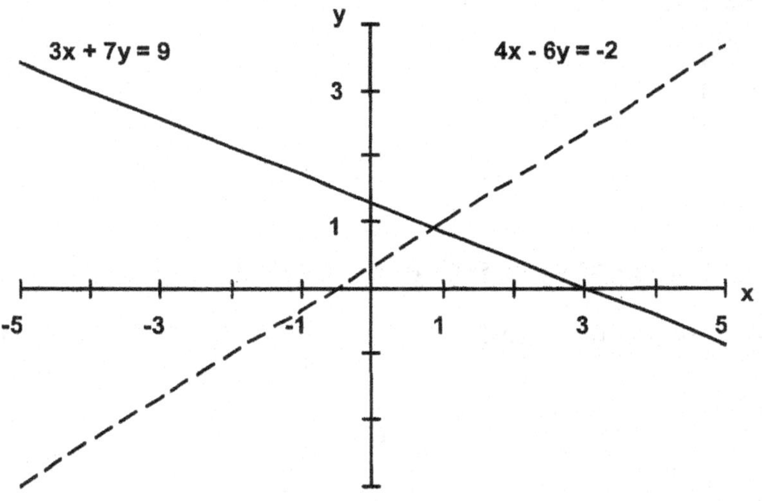

We previously studied linear functions and in this unit we will study another class of functions called **quadratic functions**. Quadratic functions have the general form,

$$F(x) = ax^2 + bx + c \,,$$

where a, b, and c are constants and $a \neq 0$.

The following example will illustrate how quadratic functions arise.

Consider a network of computers connected together to exchange email. Each pair of computers is connected directly with a separate telephone line. Suppose there were 100 computers, how many phone lines are needed? To solve this problem, let's start with a small number of computers and try and come up with a relationship that we can use to get the number of lines needed for 100 computers. We will start with a picture to see what is happening.

 Two computers, one line

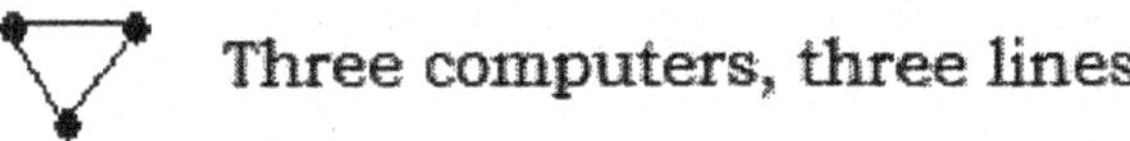 Three computers, three lines

 Four computers, six lines

Looking at the figure above, if there are four computers on the network and a new one is to be connected the new computer just needs to be connected to the old computers. That would add an additional 4 lines. Setting up a table we get

Number of computers (n)	Number of lines (t_n)
2	1
3	1 + 2 = 3
4	3 + 3 = 6
5	6 + 4 = 10
6	10 + 5 = 15
7	15 + 6 = 21

By focusing on how the number of telephone lines increases as one computer is added we can write the following, where t_n is the number of telephone lines needed for n computers:

$$t_5 = t_4 + 4$$
$$t_6 = t_5 + 5$$
$$t_7 = t_6 + 6$$

The general statement for this pattern is

$$t_{n+1} = t_n + n.$$

This relationship still does not tell us how many lines are needed for 100 computers because we would have to know how many lines are needed for 99 computers, $t_{100} = t_{99} + 99.$

From the equation $t_{n+1} = t_n + n$, let's formulate the functional equation.

$$\begin{aligned}
t_{n+1} &= t_n + n \\
&= t_{n-1} + (n-1) + n \quad \text{because } t_n = t_{n-1} + (n-1) \\
&= t_{n-2} + (n-2) + (n-1) + n \quad \text{because } t_{n-1} = t_{n-2} + (n-2)
\end{aligned}$$

Continuing this process, we eventually get to

$$\begin{aligned}
t_{n+1} &= t_1 + 1 + 2 + 3 + \ldots + n \\
&= 1 + 2 + 3 + \ldots + n, \quad \text{because } t_1 = 0
\end{aligned}$$

We know that $1 + 2 + 3 + \ldots + n = \dfrac{(n-1)n}{2}$, therefore the functional equation is

$$t(n) = \frac{(n-1)n}{2} = \frac{1}{2}n^2 - \frac{1}{2}n.$$

If we had 100 computers, we would need $t(100) = \dfrac{99(100)}{2} = 4{,}950$ telephone poles.

Note that $t(n) = \dfrac{1}{2}n^2 - \dfrac{1}{2}n$ is a quadratic function.

Activity 11.1

Which of the following functions represent quadratic functions?

a) $y = 3x - 6$

b) $y = 5 + 6x - x^2$

c) $y = -x^2 - 7x$

Section 11-2: Table of Differences

When we studied linear functions we found that when input values change at a constant rate, output values also change at a constant rate. The following table illustrates this for the linear function F(x) = 3x - 2.

x	F	1st difference
0	-2	
		1 - (-2) = 3
1	1	
		4 - 1 = 3
2	4	
		7 - 4 = 3
3	7	
		10 - 7 = 3
4	10	
		13 - 10 = 3
5	13	

Now consider the quadratic function $F(x) = 2x^2 - 3x + 2$. Look at the difference table below and notice that the first differences are not constant for this function, but the second differences are constant.

x	y	1st difference	2nd difference
0	2		
		1 - 2 = -1	
1	1		3 - (-1) = 4
		4 - 1 = 3	
2	4		7 - 3 = 4
		11 - 4 = 7	
3	11		11 - 7 = 4
		22 - 11 = 11	
4	22		

When the second difference is constant for equal differences in the independent variable, the function represents a quadratic function.

Activity 11.2

a) Complete the table of differences for the function $t(n) = \dfrac{(n-1)n}{2}$ by adding a column of first differences and another column of second differences.

n	t
0	0
1	0
2	1
3	3
4	6
5	10
6	15
7	21

b) Given the information you added to the above table, would you say the function is linear or quadratic? Justify your answer.

Activity 11.3

(The data is from *The Language of Functions and Graphs* by Shell Centre for Mathematical Education.)

Do any of the following tables represent a quadratic or a linear function?

a) **Cooling Coffee**

Time (min)	Temperature
0	90
5	79
10	70
15	62
20	55
25	49
30	44

b) **Cooking Times for Turkey**

Weight lbs	Time hrs
6	2.5
8	3.0
10	3.5
12	4.0
14	4.5
16	5.0
18	5.5
20	6.0

c) **Baby Before Birth**

Age (months)	Length (cm)
2	4
3	9
4	16
5	24
6	30
7	34
8	38
9	42

You have probably seen the quadratic formula in a previous course. This formula can be used to solve any equation of the form $ax^2 + bx + c = 0$. The formula says that solutions for x are given in terms of a, b, and c by

$$x = \frac{-b \pm \sqrt{b^2 - 4ac}}{2a}.$$

For a quadratic function $f(x) = ax^2 + bx + c$, the solutions given by the quadratic formula tell you the x-intercepts of the parabola, as long as the quantity under the square root is not negative.

The **discriminant** is the expression under the radical in the quadratic formula:

$$\text{discriminant} = b^2 - 4ac.$$

Quadratic equations can have two solutions, one solution or no solutions. The key to determining the number of solutions is the **discriminant.** If the discriminant is

> **negative** then the equation has **no** solutions,
> **zero** then the equation has **one** solution,
> **positive** then the equation has **two** solutions.

Example 11-3.1: Solve the equation $2x^2 - 8x + 3 = 0$.

Solution: Using the quadratic formula with a = 2, b = -8, and c = 3, we get

$$x = \frac{-b \pm \sqrt{b^2 - 4ac}}{2a} = \frac{-(-8) \pm \sqrt{(-8)^2 - 4 \cdot 2 \cdot 3}}{2(2)}$$

$$= \frac{8 \pm \sqrt{40}}{4} = 2 \pm \frac{\sqrt{40}}{4} \approx 2 \pm 1.58$$

Therefore, x = 2 – 1.58 = 0.42 and x = 2 + 1.58 = 3.58.   ◆

Note: We can say that the x-intercepts of the function $f(x) = 2x^2 - 8x + 3$ are (0.42, 0) and (3.58, 0).

$\boxed{\textbf{Example 11-3.2}}$: Solve $(x+1)^2 = 2(x^2 - 2x + 4)$

Solution: To use the quadratic formula we have to put the equation in the form $ax^2 + bx + c = 0$.

$x^2 + 2x + 1 = 2x^2 - 4x + 8$	Multiply $(x + 1)(x + 1)$ and distribute the 2
$x^2 - 6x + 7 = 0$	Set equation to 0
$x = \dfrac{6 \pm \sqrt{(-6)^2 - 4(1)(7)}}{2(1)} = \dfrac{6 \pm \sqrt{8}}{2}$	Solve using the quadratic equation
	with a = 1, b = -6 and c = 7
$x = \dfrac{6 + \sqrt{8}}{2}$ or $x = \dfrac{6 - \sqrt{8}}{2}$	Recall the $\pm$ means you have to evaluate a $+$ and then a -

So, $x = 4.414$ or $x = 1.586$

$\blacklozenge$

$\boxed{\textbf{Example 11-3.3}}$: Without solving the equation, determine the number of solutions that the equation $x^2 - 5x - 9 = 3$ has.

Solution: We will use the discriminant, $b^2 - 4ac$, to determine the number of solutions, but first we have to put the equation in the form $ax^2 + bx + c = 0$.

$$x^2 - 5x - 9 = 3$$
$$x^2 - 5x - 12 = 0 \quad (\textit{subtract 3 from both sides})$$

So a = 1, b = -5, and c = -12. Now substituting into $b^2 - 4ac$ we get

$$(-5)^2 - 4(1)(-12) = 73.$$

Since the discriminant is positive, the equation has two solutions. $\blacklozenge$

Activity 11.4

For each equation below, determine the number of solutions and then use the quadratic formula to solve the equation.

a) $4x^2 + 5x = 1$ b) $4x^2 - 4x = -1$

We can use the following Excel template to find the solution to a quadratic equation, but first we have to put the equation in the form $ax^2 + bx + c = 0$. If the discriminant is less than 0, a message will appear stating there is no solution.

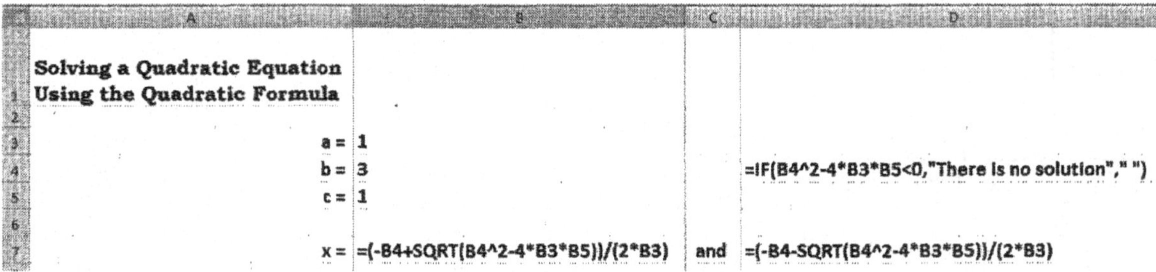

Example 11-3.4: Solve $\dfrac{x^2}{3} - 2x = -\dfrac{1}{5}$ using the quadratic formula and the Excel template from the previous page.

Solution: First let's eliminate the fractions by multiplying both sides of the equation by 15. This results in the equation

$$5x^2 - 30x = -3 .$$

Putting all terms on one side of the equation we have

$$5x^2 - 30x + 3 = 0 .$$

We will use the quadratic formula with a = 5, b = -30, and c = 3.

$$x = \frac{30 \pm \sqrt{(-30)^2 - 4(5)(3)}}{2(5)} = \frac{30 \pm \sqrt{840}}{10}$$

So, $x = 5.898$ or $x = .102$

Now use the template for Excel to verify the solutions.

Solving a Quadratic Equation Using the Quadratic Formula			
a =	5		
b =	-30		
c =	3		
x =	5.898275349	and	0.101724651

♦

Example 11-3.5: **Cox's Formula** (from *Functions and Change* by Crauder, Evans, and Noell, Houghton Mifflin Company, 2003 page 386)

Assume that a long horizontal pipe connects the bottom of a reservoir with a drainage area. Cox's formula provides a way of determining the velocity v of the water flowing through the pipe:

$$\frac{Hd}{L} = \frac{4v^2 + 5v - 2}{1200}.$$

Here H is the depth of the reservoir in feet, d is the pipe diameter in inches, L is the length of the pipe in feet, and the velocity v of the water is in feet per second.

a) Find the velocity of the water in the pipe if its diameter is 4 inches, its length is 1000 feet, and the reservoir is 50 feet deep.

b) If the water velocity is too high, there will be erosion problems. Assuming that the pipe length is 1000 feet and the reservoir is 50 feet deep; determine the largest pipe diameter that will ensure that the water velocity does not exceed 10 feet per second.

Solution: a) Here d = 4, L = 1000, and H = 50, substituting these values into Cox's formula we get

$$\frac{50(4)}{1000} = \frac{4v^2 + 5v - 2}{1200}$$

Simplifying we get

$$\frac{200(1200)}{1000} = 4v^2 + 5v - 2$$

$$240 = 4v^2 + 5v - 2$$

$$4v^2 + 5v - 242 = 0$$

Now using the quadratic formula,

$$v = \frac{-5 \pm \sqrt{5^2 - 4(4)(-242)}}{2(4)} = \frac{-5 \pm \sqrt{3897}}{8}$$

$v = \mathbf{7.178}$ **ft per second** or $v = -8.428$ ft per second

The velocity of the water in the pipe will be 7.178 ft/sec.

b) This problem is important to go through because it incorporates much of what we have covered in the course.

$L = 1000$ ft and $H = 50$ ft

$$4v^2 + 5v - 2 = 1200(\frac{50d}{1000})$$

$$4v^2 + 5v - 2 = 60d$$

$$4v^2 + 5v - 2 - 60d = 0$$

$$4v^2 + 5v - (2 + 60d) = 0$$

Now we will use the quadratic formula to solve for v, but our constant term here is $-(2 + 60d)$.

$$v = \frac{-5 \pm \sqrt{5^2 - 4(4)(-2 - 60d)}}{2(4)}$$

$$= \frac{-5 \pm \sqrt{25 + 32 + 960d}}{8}$$

$$= \frac{-5 \pm \sqrt{57 + 960d}}{8}$$

We have to find the diameter, d, such that the velocity does not exceed 10, which means we have to solve the following:

$$\frac{-5 \pm \sqrt{57 + 960d}}{8} \leq 10$$

$$-5 \pm \sqrt{57 + 960d} \leq 80$$

$$\sqrt{57 + 960d} \leq 85$$

To eliminate the square root, we can square both sides:

$$57 + 960d \leq 7225$$

$$960d \leq 7168$$

$$d \leq 7.47$$

The largest pipe has a diameter of 7.47 inches.   ♦

Name:_______________________________________   **Date:**__________

Homework Sections 11-1 – 11- 3

1. Which if the following represent a quadratic function?

   a) $g(x) = -x^2 + 4x - 10$

   b) $f(y) = 12 - 4y + y^2$

   c) $h(x) = x^3 - 7x^2 + 9$

2. Which of the following represent a quadratic function?

   a)

x	y
0	-5
1	-4
2	3
3	22
4	59
5	120

   b)

x	y
0	-7
1	-6
2	-3
3	2
4	9
5	18

3. Use the quadratic formula to solve the following equations.

   a) $x^2 + 4x - 3 = 0$ b) $x^2 = 3x + 1$ c) $y^2 + 3 = 10$

4. Find the x-intercepts of the function $f(x) = x^2 + 5x + 2$.

5. Without solving the equation, determine the number of solutions that the equation
    $2x^2 - 5x - 3 = 3$ has.

6. Solve the following using the quadratic formula.

a) $(x-2)^2 = 2x^2 + 5x$

b) $.01x^2 + .06x - .08 = 0$ (Hint: Multiply every term by 100 to eliminate the decimals.)

c) $\dfrac{1}{4}y^2 = \dfrac{2}{5}y + \dfrac{1}{10}$ (Hint: Multiply every term by 20 to eliminate the fractions.)

The graph of the quadratic function
$$y = ax^2 + bx + c, \ a \neq 0$$ is a **parabola** (U-shaped graph).

If a > 0 then the graph will open up.

If a < 0 then the graph will open down.

The **vertical intercept** is the point (0, c).

One convenient property of parabolas is that they are symmetric about a vertical line called the **line of symmetry**. The equation of the line of symmetry is $x = -\dfrac{b}{2a}$. The location of the line of symmetry is important because it determines the highest or lowest point on the curve, depending on whether the curve opens up or opens down. This **extreme point** (*minimum or maximum*) on the curve is called the **vertex** of the parabola. The x-coordinate for the vertex is $-\dfrac{b}{2a}$. By substituting $-\dfrac{b}{2a}$ for x in the equation of the curve, you can find the y-coordinate, and therefore the exact location of the vertex. We can write the vertex of a quadratic function as the point

$$\left(\frac{-b}{2a}, f\left(\frac{-b}{2a}\right) \right).$$

Example 11-4.1: Given the quadratic function $y = x^2 + 4x + 7$ determine
    i)   whether the parabola opens up or down
    ii)   the vertical intercept
    iii)   the line of symmetry
    iv)   the vertex

Solution:
    i)   Since a = 1, which is greater than 0, the parabola opens up, which means the function has a minimum.

    ii)   Since c = 7, the vertical intercept is (0, 7).

    iii)   For this function a = 1 and b = 4, so the line of symmetry, $x = \dfrac{-b}{2a}$

        is $x = \dfrac{-4}{2(1)} = -2$.

    iv)   To find the vertex we substitute x = -2 in the function to find y.

$$y = (-2)^2 + 4(-2) + 7 \ = \ 4 + -8 + 7 \ = 3$$

        So the vertex is (-2, 3).   ◆

Note: In the last example, since the parabola opens up the function has a **minimum value** which is 3 and it occurs when x = -2. The graph of the function is shown below.

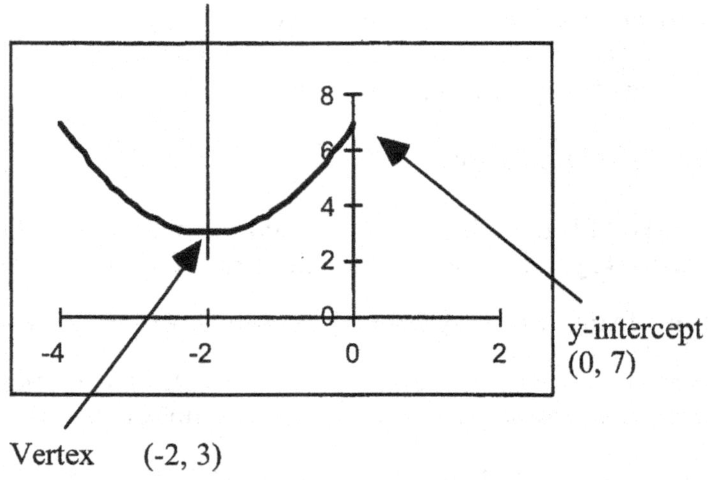

Activity 11.5

a) Given the quadratic function $y = -2x^2 - 3x - 1$ determine

  i) whether the parabola opens up or down

  ii) the vertical intercept

  iii) the line of symmetry

  iv) the vertex

  v) the maximum value of the function

b) Suppose your revenue for a product is given by the following function

$$R = -320.5p^2 + 260.3p$$, where R is the revenue and p is the price of the product in dollars.

Since the graph of the function is a parabola which opens down (a < 0),   this function has a **maximum.** Find the price that should be charged to get the maximum revenue.

We will use Excel to solve a quadratic equation, find the minimum or maximum, and given data find a quadratic model for the data.

Example 11-5.1: Solve $x^2 + 4x - 9 = 10$.

Solution: We could use the quadratic formula to solve this equation, but first we would have to subtract 10 from both sides. Remember, to use the quadratic formula, the equation has to be in the form $ax^2 + bx + c = 0$.

We will use Solver on Excel to find the solution to the equation. The equation is asking us, what value do you substitute for x to get an output of 10. Looking at the graph of $f(x)$ below we can see that there are two solutions to the equation. One is approximately 3 and the other is close to -6.

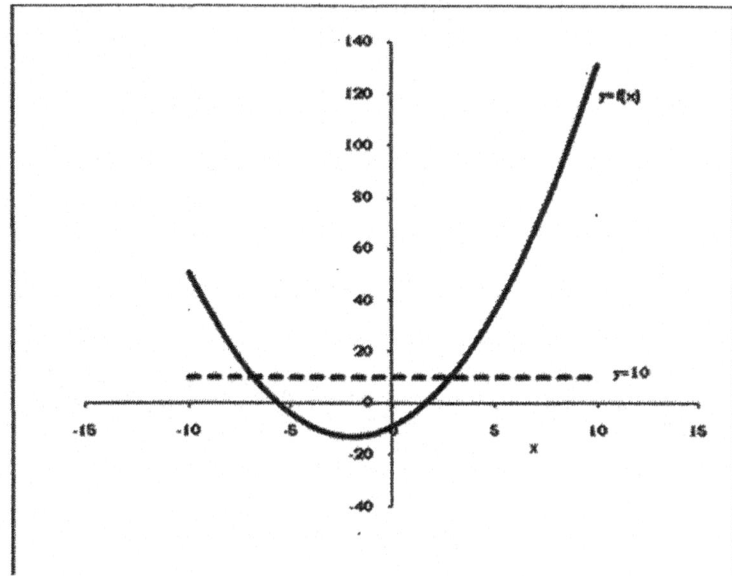

Now let's use solver.

In cell A2 enter 3, which is an initial guess for one solution.
In cell A3 enter = A2^2 + 4*A2 - 9.
Go to Data and then Solver.

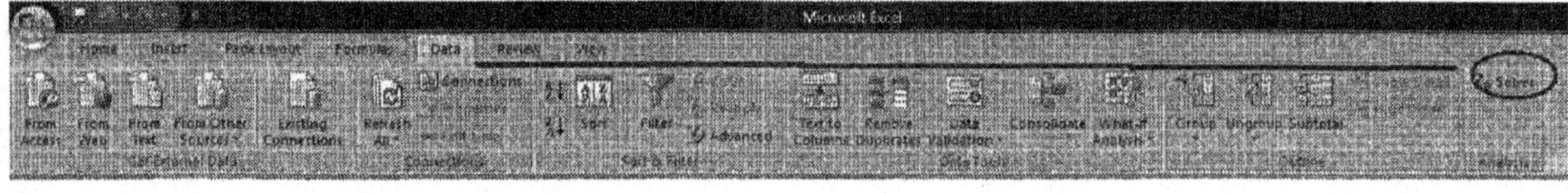

The following window will appear. Enter values as they appear in the figure on the next page.

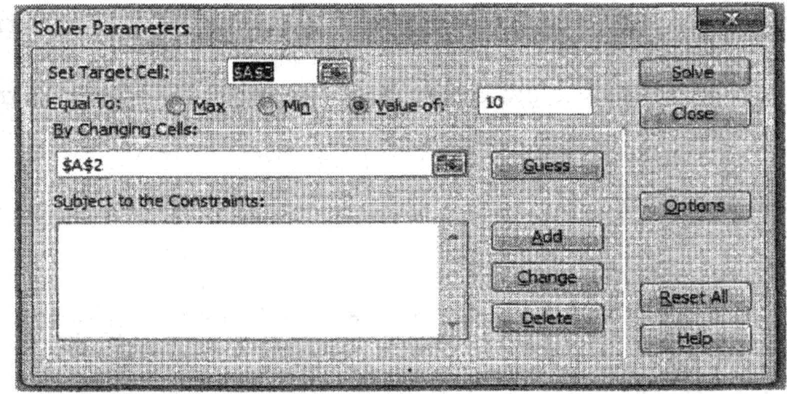

When you click Solve, one solution, 2.795831, will appear in cell A2. To approximate the second solution, the graph tells us the solution is approximately -6, so enter this in cell A2 and repeat the process. The solution -6.79583 will appear in cell A2. ♦

We could also use Goal Seek to approximate the solutions to the last equation.

Activity 11.6

Use Goal Seek to solve the equation $x^2 + 4x - 9 = 10$.

<u>*Section 11-6: Quadratic Regression*</u>

We previously looked at the following table and concluded that it represents a quadratic function. How can we determine a quadratic function that fits the data? Just as we did with linear functions, we go through the same process with Excel to find a quadratic function.

Cooling Coffee

Time (min)	Temperature
0	90
5	79
10	70
15	62
20	55
25	49
30	44

Let's first enter the data into Excel.

	A	B
1	Time	Temperature
2	0	90
3	5	79
4	10	70
5	15	62
6	20	55
7	25	49
8	30	44

To find a quadratic model that represents this data first have Excel create a scatter plot. Highlight the curve and then right click to get the dialogue box below and choose Add Trendline.

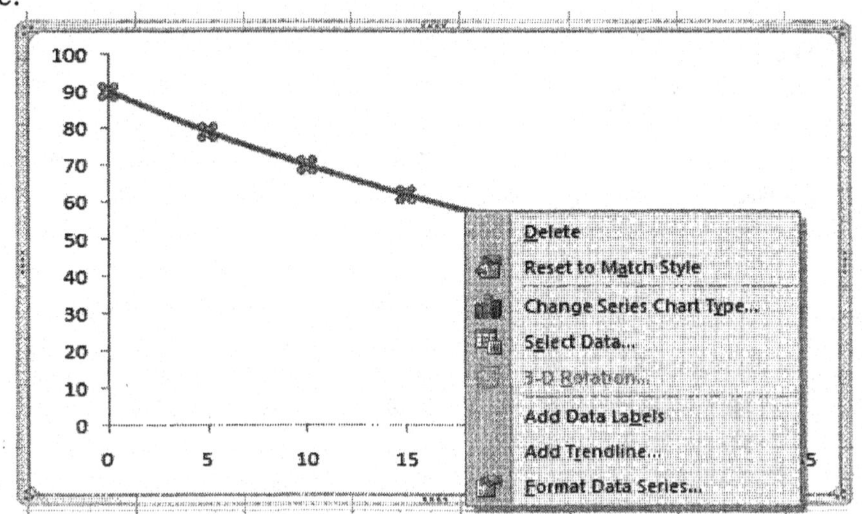

The following dialogue box appears. Choose Polynomial and order 2. In addition, click the two boxes at the bottom: Display Equation on chart and Display R-squared value on chart. Finally, close the dialogue box.

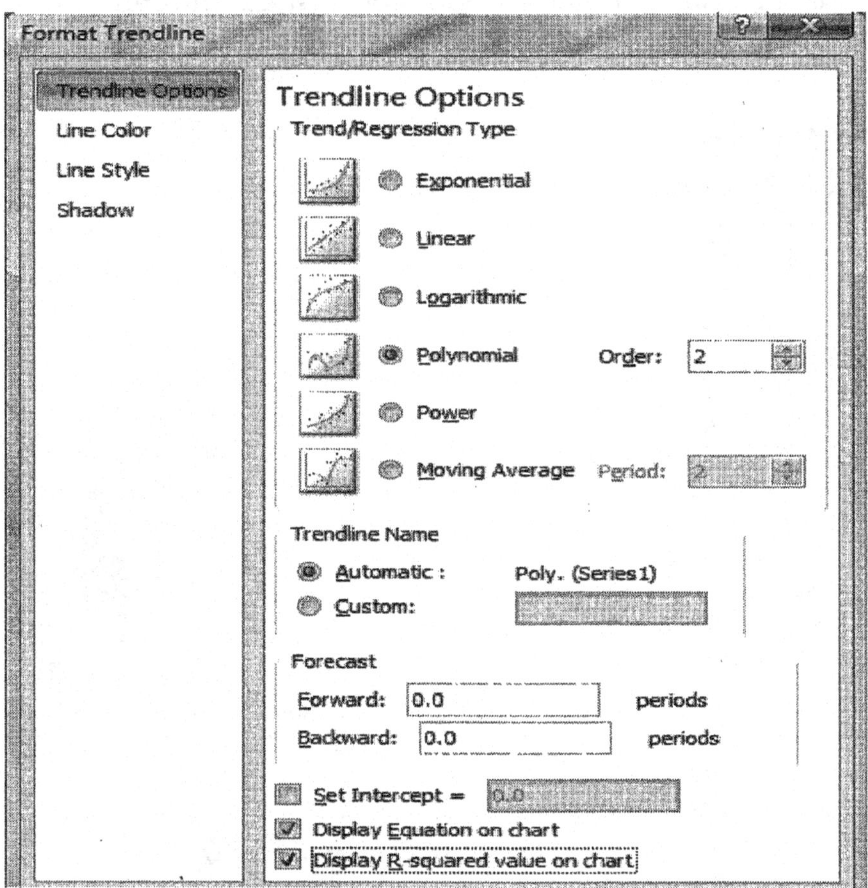

The quadratic equation and R^2 value should appear on the graph as shown below.

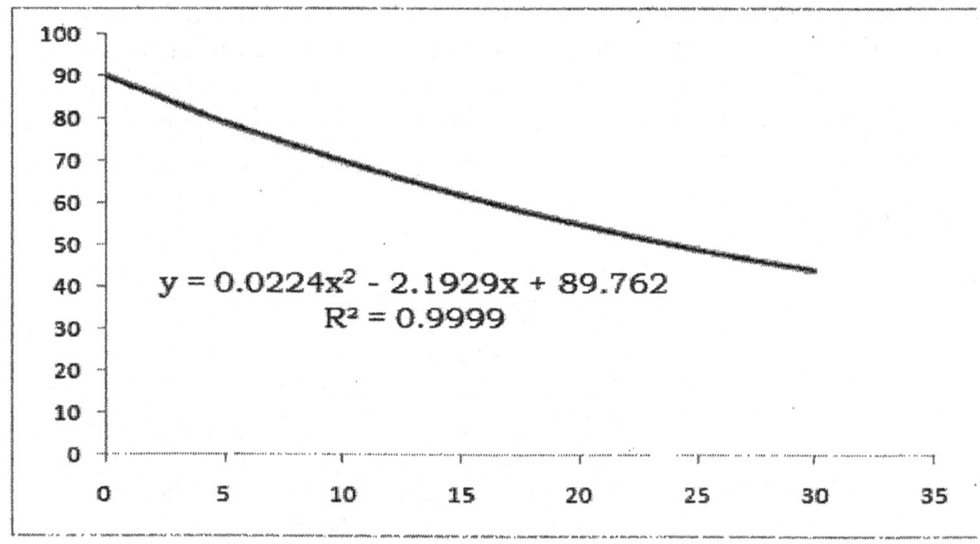

Activity 11.7

The table below shows the estimated number of AIDS cases, by year of diagnosis from 1999 to 2003 in the United States. Use Excel to find a quadratic function that models the data. (Source: US Dept. of Health and Human Services, Centers for Disease Control and Prevention, HIV/AIDS Surveillance, 2003.)

Year	AIDS Cases
0	41,356
1	41,267
2	40,833
3	41,289
4	43,171

Let year be the number of years since 1999.

Unit 11 Review Exercises

1.   a)   Show that the following data can be represented by a quadratic function.

      b)   Find a formula for the quadratic function.

x	P(x)
0	6
1	5
2	8
3	15
4	26

2.   Give an example of a quadratic function that has a graph opening down.

3.   Solve $x^2 - x = 12$ by the quadratic formula.

4.   Solve the following equations.

      a) $30x^2 + 15x = 0$

      b) $.05r^2 - .01r + .07 = .1$

      c) $x^2 - 3x = -2$

5.   Given the function $y = -3x^2 + 5x + 1$, determine

      a) whether the parabola opens up or down

      b) the vertical intercept

      c) the line of symmetry

      d) the vertex

      e) the minimum or maximum value of the function.

6.   Find the solutions to $3x^2 - 10x = 8$, using the given graph.

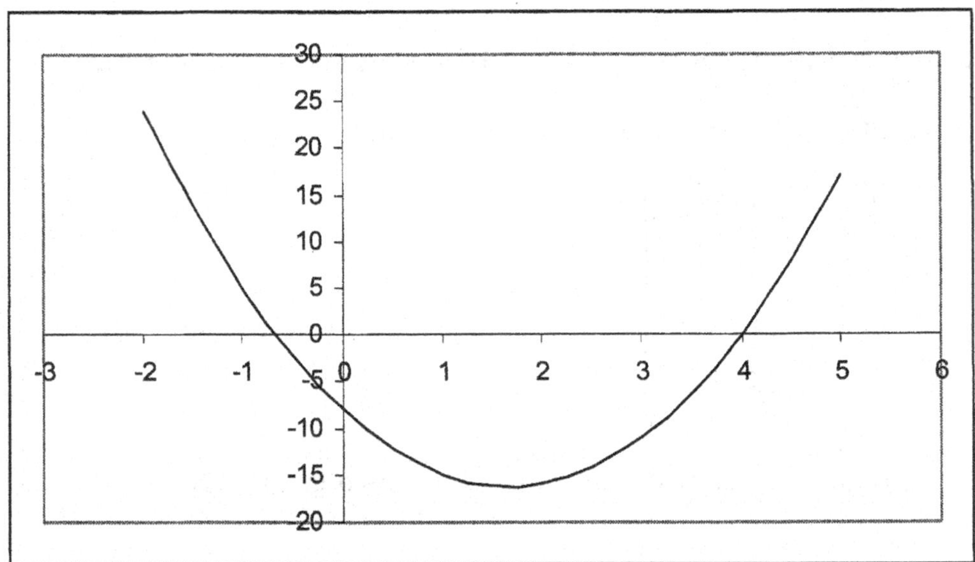

7.   Determine the number of solutions in the following quadratic equations.

a) $3x^2 - 2x - 7 = 0$ b) $4x^2 + 49 = 28x$

8.   The Total Gross State Product for New Hampshire from 1998 - 2003 is modeled by the following quadratic function:

$$TGSP = 120.88x^2 + 1430.25x + 38808.68,$$ where x = 0 corresponds to 1998 and $TGSP$ is in millions of dollars.

Using this model, predict $TGSP$ for 2005.
(Data for TGSP found at www.bea.doc.gov)

9.   The following data, which was taken from the 1950 Statistical Abstract of the United States, shows the number, in millions, of women employed outside the home in the given year. Find a quadratic function for the data.

t	Number in millions
0	16.11
1	18.7
2	19.17
3	17.53
4	13.77

Note: t represents the number of years since 1942, so t = 3 represents 1945.

10. A class of models for population growth rates in marine fisheries assumes that the harvest from fishing is proportional to the population size. One such model uses a quadratic function:

$$G = 0.3n - 0.2n^2,$$ where $\mathbf{G}$ is the growth rate of the population in millons of tons of fish per year, and $\mathbf{n}$ is the population size, in millions of tons of fish.

   a) Calculate $G(1.53)$ and explain what your answer means in practical terms.
   b) At what population size is the growth rate the largest?
   c) What population size gives a growth rate of 0.11 tons of fish per year? (Source: *Functions and Change*, by Crauder, Evans, and Noell)

11. A ball that was thrown straight up in the air that can be represented by the function $h = 96t - 16t^2$, where h is the height and t is the time the ball since the ball was thrown.

   a) Use the graph to approximate the time(s) that the ball will reach a height of 130 feet. Explain how you arrived at your estimate.

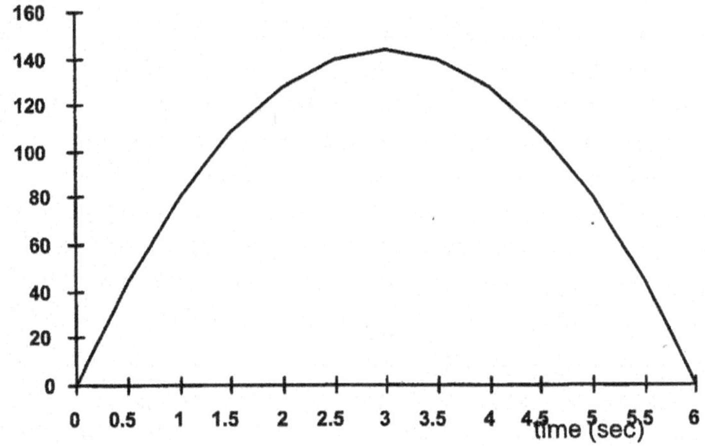

t	2	2.25	2.5	2.75	3	3.25	3.5	3.75	4	4.25	4.5	4.75	5	5.25
h	128	135	140	143	144	143	140	135	128	119	108	95	80	93

   b) Using the table, explain how you could get a more accurate estimate for the time that the ball reaches a height of 130 feet.

   c) Algebraically find the time that the ball will have a height of 130 feet. Use your calculator to approximate the solution to three decimal places. Hint: We have to solve the equation $96t - 16t^2 = 130$.

UNIT 12
Exponential Functions

Section 12-1: Introduction

Exponential functions frequently occur in nature, science, and mathematics.
Common examples are compound interest and the population growth of a
culture of bacteria. A theme that has run throughout American history as well
as world history has been growth. As Donella Meadows states in her book
Beyond the Limits,

> "Most societies, rich and poor, seek some kind of expansion as a
> remedy for their most immediate and important problems. In the rich
> world economic growth is believed to be necessary for employment, social
> mobility, and technical advance. In the poor world economic growth
> seems the only way out of poverty. And a poor family sees that many
> children can be a source not only of joy, but also of hope for economic
> security. Until other solutions are found for the legitimate problems of the
> world, people will cling to the idea that growth is the key to a better
> future, and they will go all they can to produce more growth."

Quantities can grow in two basic ways: linearly and exponentially. We have
already studied linear growth when a quantity grows by a fixed amount each
time period. Exponential growth occurs when quantities grow by a constant
multiple during a time period. It is hoped that by the end of this unit you will
realize the importance of exponential functions to real world problems.
Everyday in newspapers and on television people in authority make predictions
about growth in our economy and the American public needs the background to
be able to evaluate these predictions.

To illustrate what happens when something grows exponentially consider the
following Persian legend. A subject presented the king with a chessboard and
requested that the king give him in exchange 1 grain of rice for the first square
on the board, 2 grains for the second square, 4 grains for the third, and so
forth. The king agreed and the following table shows how many grains would
be on some of the squares.

Square	1	2	3	4	5	6	7	8	9	10	11
Grains of Rice	1	2	4	8	16	32	64	128	256	512	1024

It may not look like much exciting is happening, but in the 21st square there
would be more than one million and by the 40th there would be more than 5 x
10^{11} grains of rice. The payment could never have continued to the 64th square
because it would have taken more rice than there was in the whole world. When
something is growing exponentially many times it goes unnoticed the
suddenness with which the growth seems to expand to a huge number.

You can see how quickly an exponential function grows by looking at the
graphs on the next page of the functions

$$A(x) = 2x, \qquad \text{(linear function)}$$
$$B(x) = x^2, \qquad \text{(quadratic function)}$$
$$\text{and} \qquad C(x) = 2^x. \qquad \text{(exponential function)}$$

The first graph shows the functions for the interval 0 < x < 4. You can see that values of B(x) are greater than those of C(x). The second graph shows that when x is greater than 4, C(x) overtakes B(x) and gets much larger very quickly.

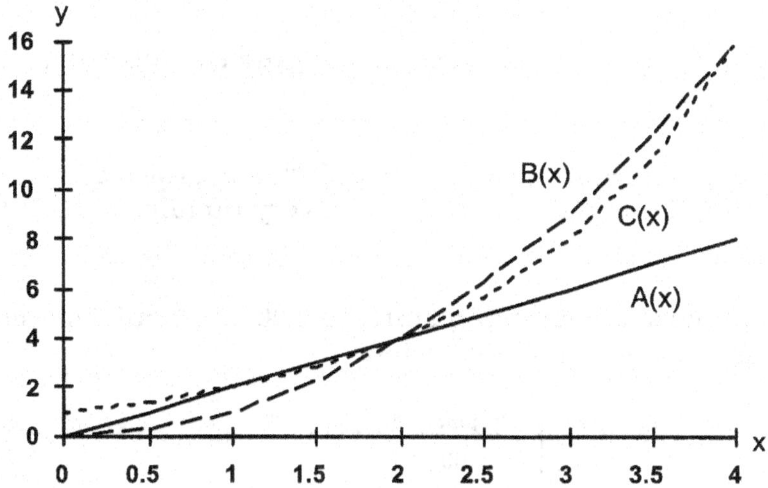

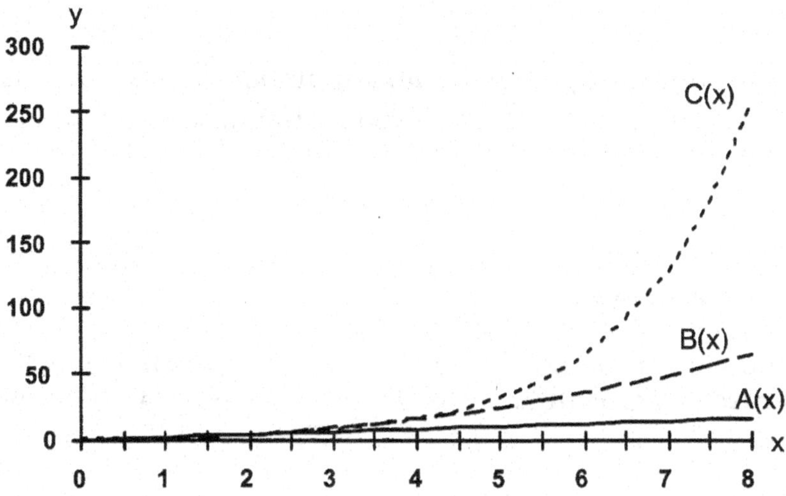

In the following sections we will explore the exponential function further so that you will recognize situations where these functions occur and use their properties to solve problems.

> The general form of an exponential function is
>
> $$P = P_0 a^t$$
>
> where P_0 is the initial quantity and **a** represents the **growth factor** if a > 1 or the **decay factor** if 0 < a < 1.

What does it mean when someone says, "the population is growing exponentially"? It means the population is growing very rapidly. Look at the following example.

Example 12-2.1: Suppose a population starts at 100 and doubles every year as shown in the table below.

Year (t)	Population (p)
0	100
1	200
2	400
3	800
4	1600

In this example the initial population, P_0, is 100 and the growth factor is 2. The exponential function then is
$P = 100(2)^t$.

Initial   Growth

Value   Factor

All exponential functions can be expressed in terms of percentages, called **growth (or decay) rates**. A growth factor of 2 is equivalent to a 100% increase.

Example 12-2.2: A growth factor of 1.05 is equivalent to what growth rate?

Solution: Suppose you started with an initial value of N and the growth rate was 1.05.

$$1.05 \cdot N = (1 + 0.05) \cdot N$$
$$= 1 \cdot N + 0.05 \cdot N$$
$$= N + (5\% \text{ of } N) \qquad (\text{Since } 0.05 \text{ is } 5\%)$$

So a growth rate of 1.05 is equivalent to a 5% increase or a growth rate of 5%.

You can see from the last example that if an amount was multiplied by 1.29, the amount will grow by 29%. It is worthwhile to make sure you understand why this is so.

$$1.29(\text{Amount}) = (1 + 0.29)(\text{Amount}) = \text{Amount} + 0.29(\text{Amount})$$

Use distributive	Original	29%
property	Amount	Increase

When the growth factor is less than 1, we have exponential **decay** rather than exponential growth.

For example, look at a table of values for the function $f(x) = 100(.8)^x$.

x	f(x)
0	100
1	80
2	64
3	51.2
4	40.96
5	32.768

You can see that this function is **decaying** (getting smaller).

Example 12-2.3: If the decay factor is 0.8, what is the decay rate?

Solution: $\quad 0.8(\text{Amount}) = (1 - 0.2)(\text{Amount}) = \text{Amount} - 0.2(\text{Amount})$

.8	Original	decaying
	Amount	20%

We can get the decay rate by rewriting 0.8 as (1 - 0.2). The decay rate would be 20%, which means that every time period the amount decreases by 20%.

♦

To see that this works, look at the table above. To go from $f(2)$ to $f(3)$ take 20% of 64 and subtract it from 64, you should get 51.2.

$$64 - 0.2(64) = 64 - 12.8 = 51.2, \text{ which is the same as } 64(0.8)$$

Example 12-2.4: Identify the following as growth or decay factors and then give the growth or decay rates.

a) 1.14   b) 1.67   c) 0.76   d) 2.1

Solution: Remember, if the factor is **greater than 1 we have growth** and if the factor is **less than 1 we have decay**.

a) Since $1.14 > 1$, 1.14 is a growth factor. To find the growth rate rewrite 1.14 as $1 + \mathbf{0.14}$, so the growth rate is **14%**. (Just change 0.14 to a percent.)

b) Since $1.67 > 1$, 1.67 is a growth factor. $1.67 = 1 + \mathbf{0.67}$, so the growth rate is **67%**.

c) Since $0.76 < 1$, 0.76 is a decay factor. $0.76 = 1 - \mathbf{0.24}$, so the decay rate is **24%**.

d) Since $2.1 > 1$, 2.1 is a growth factor. $2.1 = 1 + \mathbf{1.1}$, so the growth rate is **110%** ($1.1 = 110\%$). ♦

Activity 12.1

Identify the following as growth or decay factors and then give the growth or decay rates.

a) 1.98   b) 0.98   c) 1.33

Activity 12.2

Complete the following chart, using the first row as an example.

Function	Initial Value	Growth or Decay	Growth or Decay factor	Growth or Decay Rate
$A = 3(1.02)^t$	3	growth	1.02	2%
$B = 15(1.59)^t$				
$y = 100(.98)^x$				
$y = 1.5(2.3)^x$				

When we studied linear and quadratic functions we used a difference table to determine whether a given set of data could be represented with a linear or quadratic function. To determine whether data can be represented by an exponential function we look at ratios. The next example illustrates this.

Example 12-3.1: Determine whether the following data can be represented with an exponential function.

t	C
0	64
1	45.44
2	32.26
3	22.91
4	16.26
5	11.55

Solution: To test whether the function is exponential we will look at the ratios of the New C to the Old C.

Here are the ratios:

$$\frac{45.44}{64} = .71, \quad \frac{32.26}{45.44} = .71, \quad \frac{22.91}{32.26} = .71$$

$$\frac{16.26}{22.91} = .71, \quad \frac{11.55}{16.26} = .71$$

We can see they are all the same which tells us that the data are exponential. Keep in mind that to use this ratio method the change in the independent variable has to be constant. ♦

You want to keep in mind that with real data the ratios will not be exactly the same the way they were in the last example, but they should be close. Once we determine the data are exponential it is not hard to write the formula for the function. Simply use the general form of an **exponential function**, $P = P_0 a^t$, where P_0 is the initial quantity and a is the growth or decay factor.

Looking at Example 12-3.1, the common ratio was 0.71, which gives us the growth or decay factor. In this example since 0.71 < 1, we have decay. The initial value, value when t = 0, was 64. Hence the function is

$$C(t) = 64(0.71)^t.$$

Determine if the following data can be represented by an exponential function. If they can, write the exponential function that represents the data.

a)

x	y
0	2.6
1	7.8
2	23.4
3	70.2

b)

x	y
0	5
2	9
4	21
6	43

Homework Sections 12-1 – 12-3

1. Identify the following as growth or decay factors and then give the growth or decay rates.

    a) 1.076   b) 0.79   c) 2.12

2. The table below gives values for three functions.

    a)   Which function could be linear?

    b)   Which could be exponential?

    c)   For those which could be linear or exponential, give a possible formula for the function.

x	f(x)	g(x)	h(x)
0	16	28.4	75
1	14	25.9	60
2	11	23.4	48
3	7	20.9	38.4

3. The following table gives the population of Mexico in the early 1980's. (modified from Calculus by Deborah Hughes-Hallett page 16)

Population of Mexico (estimated), 1980 - 1986

Year	Population (millions)
1980	67.38
1981	69.13
1982	70.93
1983	72.77
1984	74.66
1985	76.60
1986	78.59

    a)   Is the population growing linearly? How can you tell?

    b)   Divide each year's population by the previous year's population. What do you notice?

c)   The fact that the common ratio is _____ shows that the population grew   by about ___% between 1980 and 1981. It is the same between any two   consecutive years. Whenever we have a constant factor, we have _____________.

d)   If we let t equal the number of years after 1980 (ie, t = 0 for 1980), the population is given by the function

initial   growth
value   factor
↓   ↓

$$P = \underline{\quad} \ (\underline{\quad})^t \ .$$

e)   Using this function, predict the population of Mexico in the year 2000.

The **federal debt** of the United States is of growing importance to anyone analyzing the functioning of the national economy. The federal debt, also called the public debt, is the total value of the government debt at a moment in time. The federal debt at the end of 2009 was about $11.9 trillion. That amounts to about $39,000 for every man, women, and child in America. The diagram below shows how the federal debt is structured. The net federal deficit is the difference between the government revenue and the government expense. When the expense is greater than the revenue the federal debt increases.

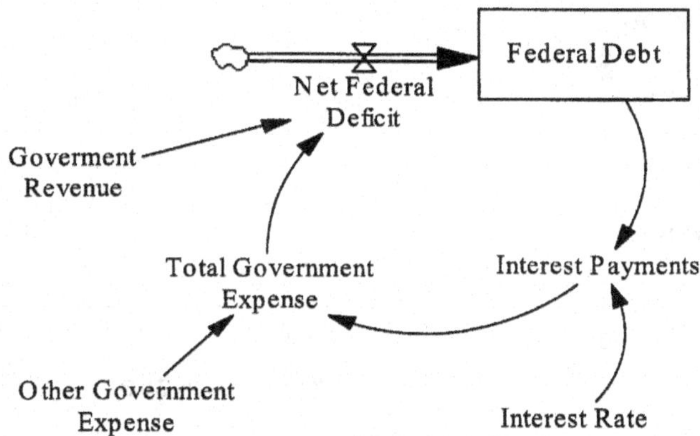

The federal debt can be modeled by the equation $\qquad d_t = d_{t-1} \cdot (1 + r_t) + (e_t - g_t)$

where
$\qquad$ d = federal debt
$\qquad$ r = interest rate
$\qquad$ e = government expense
$\qquad$ g = government revenue.

We will find functions for r, e, and g then use the equation above to predict the federal debt from 2010 through 2015. These variables are not constant and vary with time. For example, the government expenses *excluding* interest payments from 2001 to 2009 are given in the following table.

Years since 2001	Government Expense (in millions)
0	1,657,023
1	1,840,201
2	2,007,047
3	2,132,765
4	2,288,214
5	2,428,837
6	2,491,831
7	2,729,793
8	3,330,778

Let's use Excel to create a scatter plot of the data from the table.

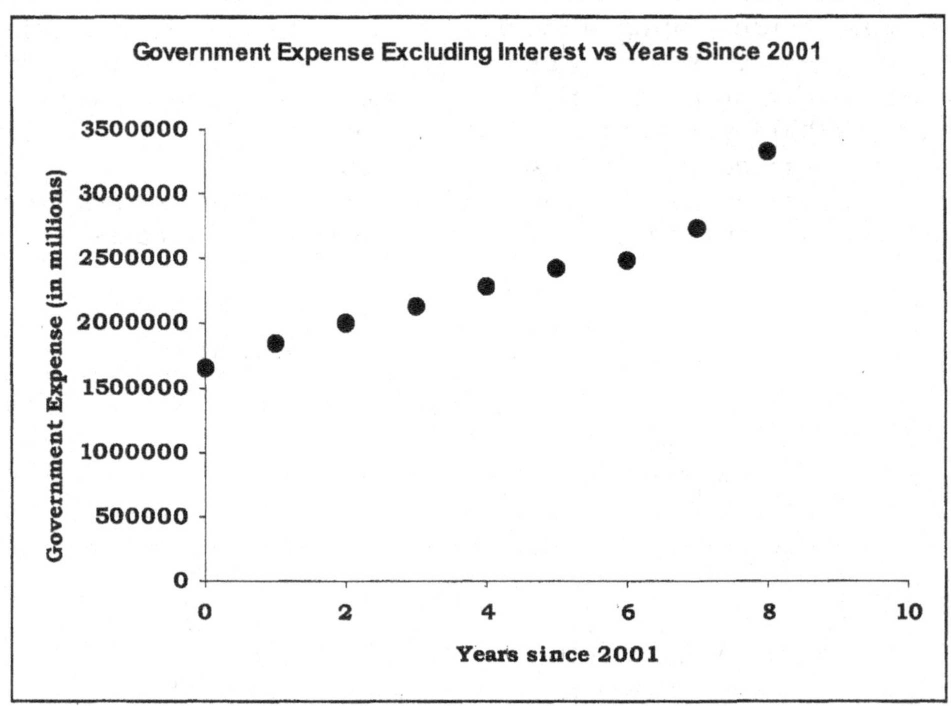

Let's fit an exponential function to this data by first finding the ratio of
government expense of one year to the preceding year. You can see on the next
page that the ratios are not all the same, but they are close, so we will use 1.09,
the average value, as the growth factor in the exponential function.

Years since 2001	Government Expense (in millions)	Ratio
0	1657023	
1	1840201	1.110546444
2	2007047	1.090667269
3	2132765	1.062638294
4	2288214	1.072886136
5	2428837	1.061455353
6	2491831	1.02593587
7	2729793	1.095496845
8	3330778	1.220157719
	Average ratio =	1.092472991

So the exponential function becomes

$$\text{Government Expense} = 1{,}657{,}023(1.092)^t$$

where t is the number of years after 2001.

Let's see how well the function fits the data. The following scatter plot shows the original data along with the values predicted by the exponential function.

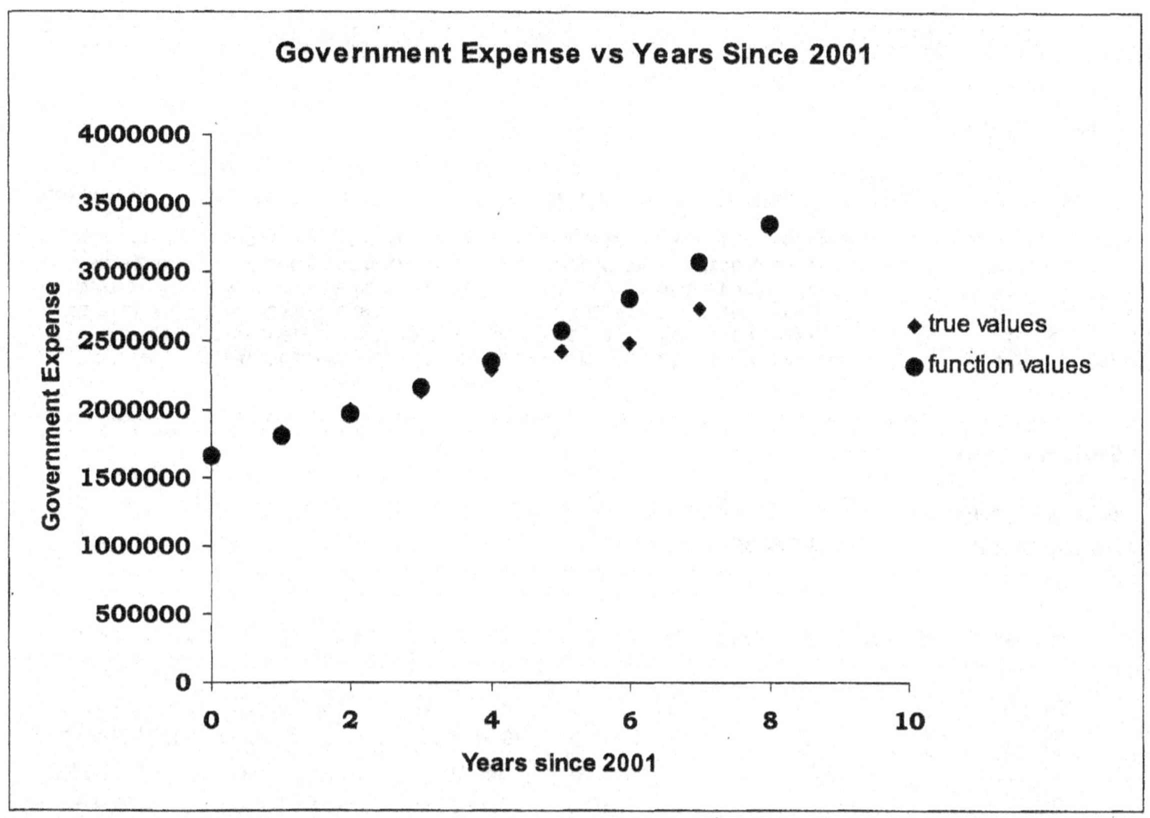

Going through the same process for government revenue and interest rate we get the following exponential functions:

$$\text{Revenue} = 1{,}991{,}430(1.011)^t$$

$$\text{Interest rate} = 0.05743(0.937)^t$$

Now along with

$$\text{Government Expense} = 1{,}657{,}023(1.092)^t$$

we can use the equation $d_t = d_{t-1} \cdot (1 + r_t) + (e_t - g_t)$ to predict the federal debt in years beyond 2009.

On the next page, we use Excel to calculate the predicted debt.

The following Excel worksheets calculate the predicted Federal Debt for the years beyond 2009. The first worksheet shows the formulas that need to be entered and the second worksheet shows the results.

	A	B	C	D	E	F
1	Federal Debt					
2						
3	Debt in 2009 (in millions)	11875851				
4						
5						
6	Year	Years after 2001	Revenue	Expenses	Interest	Federal Debt
7						11875851
8	2010	9	=1991430*(1.011)^B8	=1657023*1.092^B8	=F7*(1+0.057*0.937^B8)	=E8+(D8-C8)
9	2011	10	=1991430*(1.011)^B9	=1657023*1.092^B9	=F8*(1+0.057*0.937^B9)	=E9+(D9-C9)
10	2012	11	=1991430*(1.011)^B10	=1657023*1.092^B10	=F9*(1+0.057*0.937^B10)	=E10+(D10-C10)
11	2013	12	=1991430*(1.011)^B11	=1657023*1.092^B11	=F10*(1+0.057*0.937^B11)	=E11+(D11-C11)
12	2014	13	=1991430*(1.011)^B12	=1657023*1.092^B12	=F11*(1+0.057*0.937^B12)	=E12+(D12-C12)
13	2015	14	=1991430*(1.011)^B13	=1657023*1.092^B13	=F12*(1+0.057*0.937^B13)	=E13+(D13-C13)

	A	B	C	D	E	F
1	Federal Debt					
2						
3	Debt in 2009 (in millions)	11,875,851				
4						
5						
6	Year	Years after 2001	Revenue	Expenses	Interest	Federal Debt
7						11,875,851
8	2010	9	2197483	3658746	12252724.8	13713988.3
9	2011	10	2221655	3995351	14121776.4	15895472.4
10	2012	11	2246093	4362923	16338350	18455180
11	2013	12	2270800	4764312	18936981.6	21430493.5
12	2014	13	2295779	5202629	21954723.3	24861573.1
13	2015	14	2321033	5681271	25431419.4	28791657.5

You can see from the worksheet that the predicted Federal Debt at the end of 2010 is **$13,713,988,300,000** and at the end of 2015 the predicted Federal Debt is **$28,791,657,500,000**.

Activity 12.4

The table below gives the government revenue from 2001 to 2009. Use Excel to fit this data to an exponential function.

	A	B
1	Years since 2001	Revenue (in millions)
2	0	1,991,430
3	1	1,853,400
4	2	1,782,530
5	3	1,880,280
6	4	2,153,860
7	5	2,407,250
8	6	2,568,000
9	7	2,524,000
10	8	2,105,000

When working with exponential growth, one of the most important considerations is **doubling time**. Doubling time gives a good indication of how fast a function is increasing.

The relationship between growth rate (*as a percentage*) and the **approximate time** it will take the quantity to double is

$$\text{Doubling time} = \frac{70}{percent\ growth\ per\ unit\ time}$$

Example 12-5.1: Las Vegas has a population of 1.6 million and this last year grew at a rate of 5%. If it continued to grow at this rate, in how many years would the population double?

Solution: According to the formula above

$$\text{Doubling time} = \frac{70}{5} = 14 \text{ years}$$

The doubling time is important to know because it tells Las Vegas that in about 14 years the population will double, so the city planners have to start planning for such things as adequate schools, police service, and water. ♦

Activity 12.5

a) Suppose you invested $2,500 at an annual interest rate of 5%, how many years would it take your $2,500 to double?

b) The population of the United States is growing at the rate of 1.2% each year, if we continue to grow at this rate in how many years will the population double?

Suppose we have exponential decay rather than growth, then instead of talking about doubling time, we talk about **half-life**. The half life is the time required for the quantity to be reduced by a factor of one half.

Example 12-5.2: The release of chlorofluorocarbons destroys the ozone in the upper atmosphere. The quantity of ozone is decaying at a rate of 0.25% per year. What is the half-life of ozone?

Solution: We divide 70 by 0.25 to get the half-life.

$$\text{half-life} = \frac{70}{0.25} = 280$$

Half the present ozone will be gone in about 280 years.

♦

Name:____   Date:____________

Unit 12 Review Exercises

1. a) A growth factor of 1.78 says that something grew by ____% each period.

    b) A decay factor of 0.85 says that something decreased by ___% each period.

    c) Write an exponential function that has an initial value of 275 and a growth rate of 3%.

2. A colony of bacteria has an hourly growth factor of 1.3. The size of the colony is determined by its weight W in milligrams.

    a) If the weight at noon is 2.0 milligrams, complete the following table.

Time (T) number of hrs since noon	Weight (W) in mg
0	2.0
1	
2	
3	
4	

    b) When T = 2, what time is it?

    c) Write the exponential function for this data, W as a function of T.

    d) What is the weight of the colony at the end of 10 hours?

    e) How much did the colony weigh at 11 am, one hour before it weighed 2.0 mg? Explain how you got your answer.

3. Suppose at birth you were given a $100 savings bond, which earns interest at 7.5% each year. This means the growth factor is 1.075 each year.

    a) Find the value of the bond at the end of 1 year, 2 years, and 3 years.

    b) Write V, the value of the bond, as a function of T, measured in years after birth.

    c) What will be the value of the bond when you are entering college, T = 18?

4.   Given the following data, how do you know it represents an exponential function? What is the formula for the exponential function?

T	N
0	120.00
1	124.80
2	129.79
3	134.98
4	140.38

5.   Given the following data, what type of function does the data represent? Write a formula for the function.

F	G
0	-4.10
2	-0.10
4	3.90
6	7.90
8	11.90

6.   The frets on a guitar are placed so that they make the correct vibrating string length for the note of music. We are interested in how the vibrating string length changes for each fret position, so we need to measure the length from the bridge to each fret position. The table below gives the lengths in order from the longest (fret 0) to fret 12. The lengths are in millimeters.

Fret number	0	1	2	3	4	5	6	7	8	9	10	11	12
Length	660	623	588	555	524	494	467	440	416	392	370	350	330

   a)   Find the ratio for this exponential decay function.

   b)   Find the decay rate.

   c)   Find a formula for the function.

   d)   At what fret does the vibrating string length reach its half-length?

7.   If inflation was 9.5% per year and if it continued at this pace, how long would it take prices to double?

8.   If the inflation rate is 1.1% per month, what is the inflation rate per year?

9.   Suppose that the population of a particular species is initially 800 and is declining at a rate of 10% per year. What will the population be 7 years from now?

10.   A population is growing exponentially according to the graph below. Write a possible formula for the function. Estimate the doubling time of the population from the graph.

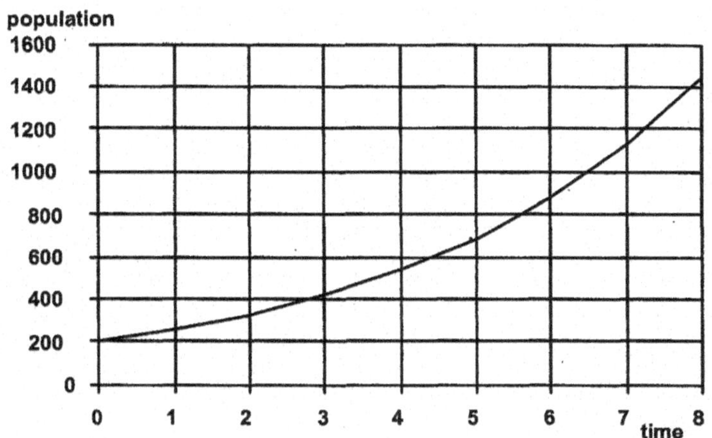

11.   Suppose we wanted to remove pollutants from water by passing it through a pipe that filtered the water. For each inch of pipe the water passes through, 70% of the pollutants in the water are removed. If there are initially 10 units of the pollutant in the water and **n** is the number of inches of pipe the water passes through, answer the following questions.

a)   Write the exponential function for the pollutants removed.

The exponential function is P = (initial value)(decay factor)^n .

P = ________________

b)   Use Excel to graph the function.

c)   How many units of the pollutant are in the water after the water passes through 10 inches of pipe?

X Y
Z

UNIT 13
Logarithms

Section 13-1: Introduction

In the last unit, we projected the population of Mexico (in millions) by the function

$$P = f(t) = 67.38(1.02)^t ,$$

where t is the number of years since 1980. Suppose that instead of calculating the population at time t, we were asked when the population will reach 100 million, in other words, we want to find the value of t such that

$$100 = 67.38(1.02)^t .$$

Our unknown here is an exponent and we could use Solver or Goal Seek to solve the equation. However, in this unit we will use **logarithms** to solve for a variable that is an exponent.

In addition to solving exponential equations, logarithms are applied in logarithmic scales and also for transforming data. Some examples of logarithmic scales are the pH scale in chemistry, decibel scale used to measure sound intensity, and the Richter scale used to measure earthquake strength.

When we transform data, we are trying to represent in a linear form a relationship that is originally nonlinear. If the graph of the transformed data falls on or near a straight line, we can use the equation of that straight line to find an equation relating the original variables.

Section 13-2: Definition of Logarithms

For any base b, we define the logarithm base b as follows.

Logarithm base b

$$y = \log_b x \quad \text{means} \quad b^y = x \quad \text{where } x > 0, \, b > 0, \text{ and } b \neq 1$$

You can see from the definition, ***logarithms are exponents***. So, $\log_3 9$ is really asking you for the power of 3 needed to get 9 and since $3^2 = 9$, $\log_3 9 = 2.$

Example 13-2.1: Using the definition above, find the following.

$$\text{a) } \log_4 16 \qquad \text{b) } \log_2 8 \qquad \text{c) } \log_3 1$$

Solution: a) $\log_4 16$ is the exponent y such that $4^y = 16$ and since $4^2 = 16$,

$$\log_4 16 = 2.$$

---

b) We are looking for the exponent y such that $2^y = 8$. Since
$2^3 = 8$, $\log_2 8 = 3$.

c) Since $3^0 = 1$, $\log_3 1 = 0$.   ◆

Activity 13.1

Find $\log_2 (1/4)$.

Your calculator has two buttons to calculate logarithms. The **LOG** button calculates logarithms with base 10. For example, if you press **LOG 100** your calculator will give you 2, since $10^2 = 100$. Logarithms base 10 are called *common logarithms*. When you see a logarithm written without the base, it is understood to be a common logarithm. For example, log 100 is understood to have base 10.

Example 13-2.2: If you press **LOG 40** you get 1.60206, since $10^{1.60206} = 40$.
If you check, $10^{1.60206}$ should give you something close to 40. It will not be exact, since we did some rounding.   ◆

The second logarithm button on your calculator is labeled **LN**. This calculates logarithms with base e, where $e \approx 2.718281828$. The logarithm base e is called the *natural logarithm*.

Example 13-2.3: If you press **LN 5** you get 1.6094379. Remember in exponential form, this means $e^{1.6094379} = 5$.   ◆

What would you do if you had to calculate $\log_2 9$? You don't have a logarithm base 2 button on your calculator. You can calculate this a couple ways. One method is to use the following formula:

$$\log_b x = \frac{\log x}{\log b}.$$

Using this property to find $\log_2 9$, we get

$$\log_2 9 = \frac{\log 9}{\log 2} \approx \frac{0.95424}{0.30103} \approx 3.17.$$

You could also use Excel to calculate a logarithm to any base. To calculate $\log_2 9$ you would enter **=log(9,2)**.

Note: If you entered **=log(9)**, Excel interprets this as a common logarithm. In addition if you wanted to calculate ln 5 on Excel, you would enter **=ln(5)**.

Activity 13.2

Use Excel or your calculator to find the following:

a) log 50   b) $\log_5(3/8)$ c) ln 35

Section 13-3: Properties of Logarithms

To work with logarithms, we use the following properties:

Properties of Logarithms

There are three basic properties of logarithms:

$$\text{I.} \qquad \log_b (xy) = \log_b x + \log_b y.$$

$$\text{II.} \qquad \log_b \left(\frac{x}{y}\right) = \log_b x - \log_b y.$$

$$\text{III.} \qquad \log_b x^n = n \cdot \log_b x.$$

Example 13-3.1: Use the properties of logarithms to rewrite each expression.

a) $\log_b 6x$ b) $\log_b\left(\dfrac{b^3}{6}\right)$ c) $\log_5\left(\dfrac{25}{x^5}\right)$ d) $\log_b\left(\dfrac{x^2 y^3}{5}\right)$

Solution:

a) $\log_b 6x = \log_b 6 + \log_b x$ Property 1

b)
$$\log_b\left(\frac{b^3}{6}\right) = \log_b b^3 - \log_b 6 \qquad \text{Property 2}$$
$$= 3 \log_b b - \log_b 6 \qquad \text{Property 3}$$
$$= 3 - \log_b 6 \qquad \text{since } \log_b b = 1$$

c)
$$\log_5\left(\frac{25}{x^5}\right) = \log_5 25 - \log_5 x^5 \qquad \text{Property 2}$$
$$= 2 - 5 \log_5 x \qquad \text{Property 3 and } \log_5 25 = 2$$

d)
$$\log_b\left(\frac{x^2 y^3}{5}\right) = \log_b x^2 y^3 - \log_b 5 \qquad \text{Property 2}$$
$$= \log_b x^2 + \log_b y^3 - \log_b 5 \qquad \text{Property 1}$$
$$= 2 \log_b x + 3 \log_b y - \log_b 5 \qquad \text{Property 3}$$

♦

Activity 13.3

Express the following as a single logarithm. (Do the reverse of what was done in Example 13-3.1.)

a) $\quad \dfrac{1}{2}\log_b x \ - \ \log_b(1+x^2)$
b) $\quad 2\log_b x \ - \ 3\log_b y$

Example 13-3.2: Evaluate 253^{65} using your calculator.

Solution: You can use the properties of logarithms and exponents to calculate this. One important property of logarithms that we did not mention is when two numbers are equal the logarithms of the numbers are equal. In other words,

If x = y, then $\log_b$ x = $\log_b$ y.   **Property IV**

Let's try to use this to evaluate 253^{65}.

Suppose $253^{65} = x$ We want to find what x is equal to.

$\log 253^{65} = \log x$ By property IV. It does not matter what base we use. Base 10 is convenient because we can use a calculator with base 10 logarithms.

$65 \log 253 \ = \ \log x$ Rewrite the left side using property III.

$156.20 \quad = \ \log x$ Evaluate log 253 with your calculator and multiply by 65.

$10^{156.02} \ = \ x$ Write the last equation in exponential form. ($y = \log x$ means $10^y = x$)

$10^{.02} \cdot 10^{156} \ = \ x$ Use properties of exponents to simplify.

$1.047 \cdot 10^{156} \ = \ x$ Evaluate $10^{.02}$

♦

Note: The 4 properties of common logarithms are true for natural logarithms as well.

Homework Sections 13-1 – 13-3

1.   Write $5^3 = 125$ in log form.

2.   Without a calculator or Excel, find the following.

   a) $\log_2 32$ b) $\log_5\left(\dfrac{1}{25}\right)$ c) $\log 1000$

3.   Use the properties of logarithms to rewrite each expression.

   a) $\log_b b^5$ b) $\log_b\left(\dfrac{7x}{6}\right)$ c) $\log_3\left(\dfrac{27}{x^6}\right)$ d) $\log_b\left(\dfrac{x^2}{2}\right)^3$

4.   Express the following as a single logarithm.

   a) $\log_2 x - \log_2(x-7)$

   b) $\dfrac{1}{3}\log_b x + 3\log_b y$

Now that we know the properties of logarithms, we can solve an equation like $5^x = 20$ algebraically. Look at the following example.

Example 13-4.1: Solve the following:
$$5^x = 20$$

Solution: We will first use property IV, taking the logarithm of both sides.

$$\log 5^x = \log 20 \qquad \text{property IV}$$

$$x \log 5 = \log 20 \qquad \text{property III}$$

$$x = \frac{\log 20}{\log 5} \qquad \text{divide both sides by log 5}$$

Using your calculator enter the following:

Scientific calculator: [**20**] [**LOG**] [÷] [**5**] [**LOG**] [=]

or depending on your calculator

[**LOG**] [**20**] [÷] [**LOG**] [**5**] [=]

Graphing calculator: [**LOG**] [**20**] [÷] [**LOG**] [**5**] [**ENTER**]

x = 1.86   (**Check the result on your calculator. Does $5^{1.86} = 20$?**)

♦

Note: $\dfrac{\log 20}{\log 5}$ is not the same as $\log \dfrac{20}{5}$.

Activity 13.4

Solve the following.

a) $4^x = 105.8$ b) $254.7(1.025)^x = 290$

Example 13-4.2: Solve $1.9 = e^{0.04t}$ for t.

Solution: Since we have base e, we will start by taking the natural logarithm of both sides.

$$\ln 1.9 = \ln e^{0.04t}$$

Using property III, we get

$$\ln 1.9 = 0.04t(\ln e)$$

Since $\ln e = 1$, we have

$$\ln 1.9 = 0.04t$$

Solving for t, we get

$$t = \frac{\ln 1.9}{0.04} = 16.046$$

Make sure you check your solution. Does $e^{0.04\,(16.046)} = 1.9$?

Activity 13.5

Solve the following.

a) $e^x = 7.2$

b) $45 = 20e^{0.025t}$

Another way to solve an exponential equation is by looking at the graph. This solution, however, will usually be an approximation. Let's look at the equation from example 13-4.2, but give it some context.

Example 13-4.3: The graph below represents the function $B(t) = e^{0.04t}$, where B(t) represents the balance (in thousands) in a bank account and t represents the years after 1995. Determine when the balance will reach $1900, or in other words, approximate the solution to $1.9 = e^{0.04t}$.

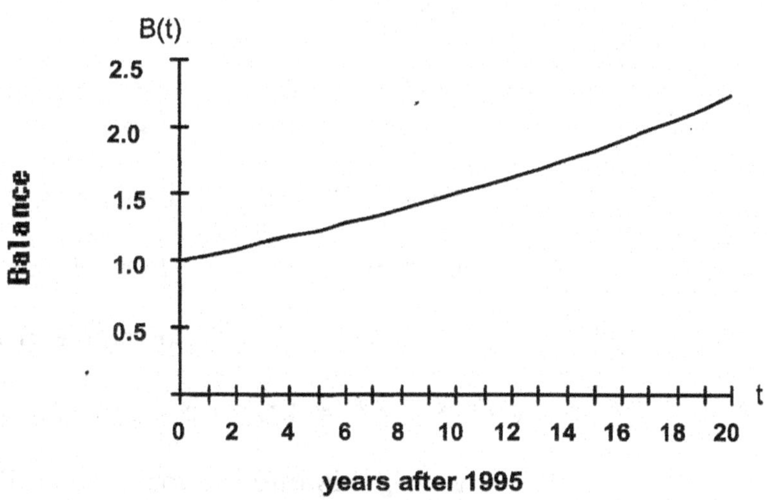

Solution: In order to find the solution to $1.9 = e^{0.04t}$, we will need to locate 1.9 on the vertical axis. Once we are there, we need to go over to where 1.9 hits the graph and then drop down to read the value on the horizontal axis. When we do that, we find that t is approximately 16 years after 1995. ◆

Activity 13.6

Use the graph of the function, $f(x) = 3^x$, to solve the equation $3^x = 7$.

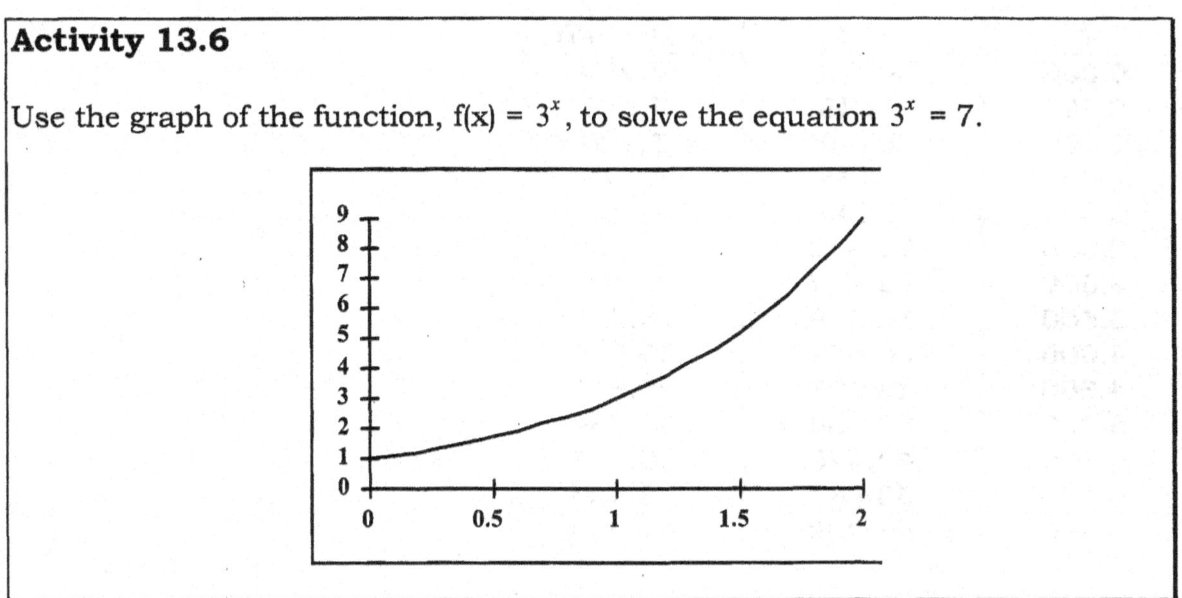

Sometimes we may have to change the base of an exponential function.

Example 13-5.1: Rewrite the following exponential function so the base is e.
$$P(t) = 3.2(1.6)^t$$

Solution: Since 1.6 is the base, we have to solve the equation

$$e^x = 1.6.$$

$$\ln e^x = \ln 1.6 \qquad \text{(Take the natural logarithm of both sides.)}$$

$$x \ln e = \ln 1.6 \qquad \text{(Use property III)}$$

$$x = \ln 1.6 \qquad \text{(ln e = 1 and } x \cdot 1 = x\text{)}$$

$$x = 0.47 \qquad \text{(Evaluate ln 1.6)}$$

Therefore, $e^{0.47} = 1.6.$ (Check this on your calculator.)

Substituting this into the formula for P(t) we get

$$P(t) = 3.2(e^{0.47})^t = 3.2e^{0.47t}.$$

You can see by examining the next table that the two functions are equal.

t	$3.2(1.6)^t$	$3.2(e^{.47t})$
0.000	3.200	3.200
0.500	4.048	4.048
1.000	5.120	5.120
1.500	6.476	6.476
2.000	8.192	8.192
2.500	10.362	10.362
3.000	13.107	13.107
3.500	16.579	16.579
4.000	20.972	20.971
4.500	26.527	26.527
5.000	33.554	33.554
5.500	42.443	42.443
6.000	53.687	53.686
6.500	67.909	67.908

♦

In the last unit we came up with the exponential function $P(t) = 67.38(1.026)^t$ for the population of Mexico for any number of years after 1980. Let's try to rewrite this function so it has a base of e rather than 1.026. What we are trying to solve is

$$1.026 = e^x.$$

Let's solve it the way we did the last example.

$$1.026 = e^x$$

_____________   Take the natural logarithm of both sides. Why is this convenient?

_____________   Use property III of logarithms.

Now replace 1.026 with $e^{.0257}$ in the function P(t).

Unit 13 Review Exercises

1. Use the method of example 13-4.1 to solve the $5^x = 33$.

2. Use the graph below of the function $f(x) = 5^x$ to solve the equation
    $5^x = 33$.

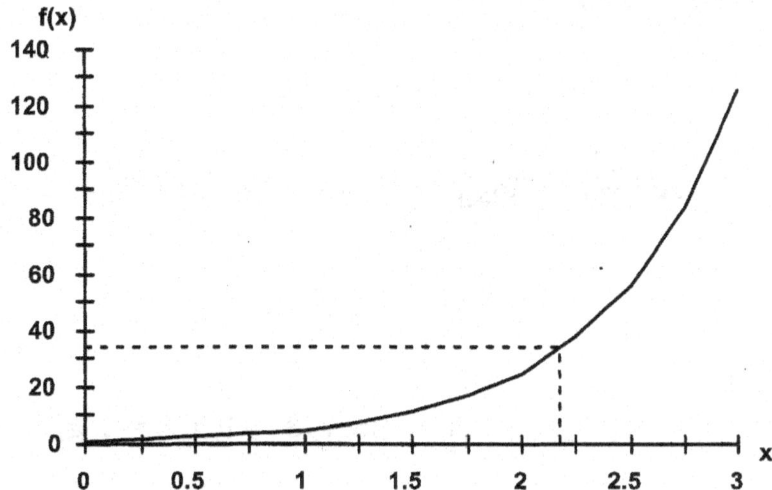

3. Write the following in exponential form.

    a) $\log 100 = 2$ b) $\log_6 6 = 1$

    c) $\log_7 343 = 3$

4. Find the value of the following:

    a) $\log_2 8$ (Remember this is asking, "*What power do you raise 2 to in order to get 8?*".)

    b) $\log_3 27$ c) $\log_4 64$

    d) $\log_4 \dfrac{1}{16}$ e)   ln e   (e to what power gives you e)

5. Use your calculator to find the following.

    a) $e^t = 34.5$ b) $10^t = 150$

6.   Apply property I, $\log_b (xy) = \log_b x + \log_b y$ to the following:

a) $\log_b 6 = \log_b(3 \cdot 2) = $ ______ + ______

b) $\log_b 42 = $ ________ = ______ + ______

c) $\log_b 5x = $ ________ = ______ + ______

d) $\log_b 5 + \log_b 7 = \log_b(5 \cdot 7) = $ ________

e) $\log_b 2 + \log_b 5 + \log_b 7 = $ ______________

7.   Apply property II, $\log \left(\dfrac{x}{y}\right) = \log_b x - \log_b y$ to the following:

a) $\log_b \dfrac{9}{2} = $ ____ - ____   b) $\log_b \dfrac{19}{15} = $ ____ - ____

c) $\log_b \dfrac{5}{33} = $ ____ - ____   d) $\log_b 13 - \log_b 3 = \log_b$ ——

e) $\log_b 35 - \log_b 7 = $ ________

8.   Apply property III, $\log_b x^n = n \cdot \log_b x$ to the following:

a) $\log_b 5^t = $ ____ $\log_b$ ____

b) $\log_b 81 = \log_b 9^2 = $ __________

c) $\log_b 8 = $ ________ = ________

d) $4 \log_b 3 = \log_b$ ____ = ________

9.   Solve the following equations.

a) $7^x = 21$ b) $4.5(1.05)^x = 33$ c) $3e^{0.5x} = 12$

10.   Rewrite the following function so the base is 10.

     $Q(t) = 4.5(1.3)^t$.   (Note: We have to first solve the equation $10^x = 1.3$.)

UNIT 14

Arithmetic and Geometric Sequences

Section 14-1: Introduction

In this unit we will study *arithmetic* and *geometric* sequences. Examples of these
are all around us in applications that have a discrete and predictable pattern.
As seen in an earlier unit, simple interest is an example of an arithmetic
sequence and compound interest is an example of a geometric sequence. We
have also looked at automobile depreciation. Since depreciation is fairly
constant for a particular model, a car's yearly value can be determined using a
geometric sequence. The twelve tones in an octave form a geometric sequence
so that the end of an octave has a frequency twice that of its low tone.

Section 14-2: Arithmetic Sequence

Definition: An *arithmetic sequence* is a sequence of numbers where each term
is a sum of the preceding one and a fixed number. This fixed number is called
the common difference.

Example 14-2.1: The following are examples of arithmetic sequences:

a)   2, 5, 8, 11, 14, ...
     Here the initial term is 2 and the common difference is 3. Notice
     the difference between terms is the common difference.

b)   10, 9, 8, 7, 6, ...
     Here the initial term is 10 and the common difference is -1.

c)   5, 7, 9, 11, ...
     Here the initial term is 5 and the common difference is 2.   ♦

We can use Excel to generate an arithmetic sequence. The worksheet below will
generate an arithmetic sequence with initial term 2 and common difference 3.

	A	B	C
1		initial term =	2
2		difference =	3
3			
4	**term**	**Arithmetic Sequence**	
5	1	=C1	
6	2	=B5+$C$2	
7	3	=B6+$C$2	
8	4	=B7+$C$2	
9	5	=B8+$C$2	
10	6	=B9+$C$2	
11	7	=B10+$C$2	
12	8	=B11+$C$2	

We defined the sequence recursively in cells B6 through B12.

> **Definition**: A *recursive definition* for a sequence is a set of statements that specifies one or more initial terms and defines the nth term, a_n, in relation to one or more of the preceding terms.

Look at the following sequence:

Term	1	2	3	4	5	6	7	8	...
Sequence	**2**	**5**	**8**	**11**	**14**	**17**	**20**	**23**	...

Using subscript notation, a_{10}, represents the 10th term in the sequence. Our sequence above would be defined recursively as

$$a_1 = 2, \quad a_n = a_{n-1} + 3.$$

You read this recursive formula as, the initial term of the sequence is 2 and to get any term in the sequence add 3 to the preceding term. Subscript notation will take time to get used to if you haven't seen it before.

It may help to write out the terms of an arbitrary arithmetic sequence.

$$a_1, a_2, a_3, \ldots, a_{n-2}, a_{n-1}, a_n, a_{n+1}, \ldots$$

Activity 14.1

Suppose you have an arithmetic sequence where the common difference, d, is 4 and $a_{20} = 66$, find

a) $\qquad a_{22}$

b) $\qquad a_{17}$

Example 14-2.2: In the sequence of figures shown below make some observations about the pattern suggested and write a recursive formula for the number of squares in the nth figure.

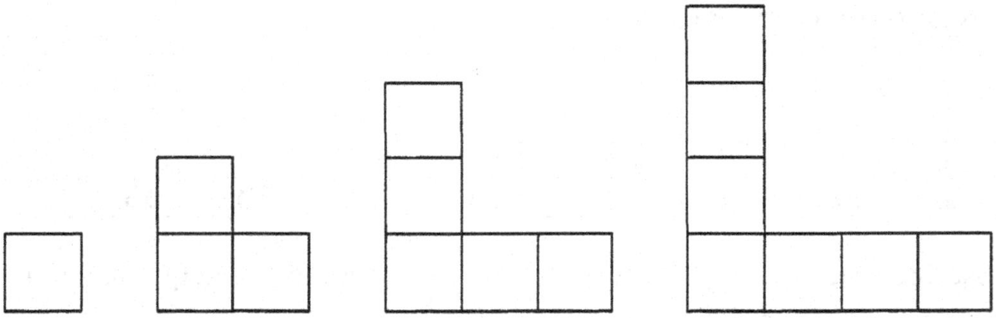

Figure 1   **Figure 2**   **Figure 3**   **Figure 4**

Solution: One observation is that the each figure has two more squares than the figure that precedes it. For example, Figure 2 has 3 squares and Figure 3 has 5 squares. If we let a_3 represent the number of squares in Figure 3, we can write the following formulas:

$$a_2 = a_1 + 2$$
$$a_3 = a_2 + 2$$
$$a_4 = a_3 + 2$$

Generalizing this pattern we can write

$$a_n = a_{n-1} + 2.$$ ◆

When using a recursive formula, we would need to know the 999th term to find the 1000th term of an arithmetic sequence. It would be easier if we could find a formula that would give us any term of a sequence in terms of just the initial value. For example, in the above problem an explicit formula for the sequence of squares is $a_n = 1 + 2(n-1)$. Let's look at the next activity to see how we can develop an explicit formula.

Activity 14.2

The first term of an arithmetic sequence is 5; its common difference is 3. Find the 100th term of the sequence. Let's start by out writing the terms.

1st term	5
2nd term	$5 + 3 = 8$
3rd term	$(5 + 3) + 3 = 5 + 2(3) = 11$
4th term	$(5 + 2(3)) + 3 = 5 + 3(3) = 14$
5th term	$(5 + 3(3)) + 3 = 5 + 4(3) = 17$

Looking for a pattern in the first 5 terms, what would the 100th term be?

In general then, if the first term of an arithmetic sequence was a_1 and its common difference was d, what would the nth term be?

Definition: The explicit formula for a sequence defines a_n with an expression that does not involve previous terms. In general, the expression for the **nth term** of an *arithmetic sequence* where d is the common difference and a_1 is the first term is:

$$a_n = a_1 + (n-1)d .$$

Activity 14.3

In an arithmetic sequence whose difference is d every other term is deleted. Do you get an arithmetic sequence? If so, what is the difference?

| **Example 14-2.3**: | The first term of an arithmetic sequence is 5, the third term is 9. What is the second term? |

Solution: **Method 1** The sequence looks like 5, s, 9, where s is the second term. This is an arithmetic sequence, so we know there is a common difference, d. This allows us to write

$$s = 5 + d, \text{ and}$$
$$s = 9 - d.$$

Since $5 + d$ and $9 - d$ are equal to the same thing, we can write

$$5 + d = 9 - d.$$

Solving the equation for d,

$$5 + d = 9 - d$$
$$5 + 2d = 9 \quad \text{(Add } d \text{ to both sides)}$$
$$2d = 4 \quad \text{(Subtract 5 from both sides)}$$
$$d = 2 \quad \text{(Divide both sides by 2)}$$

Since $d = 2$, the second term is $5 + 2 = 7$.

Method 2 We know the sequence is arithmetic so visualizing the sequence we have

$$5, \underbrace{\quad}_{d} s, \underbrace{\quad}_{d} 9.$$

The difference between 9 and 5 is 4, and this has to be divided between two ds, so d has to equal 4/2. Therefore the second term s has to be $5 + 2 = 7$.

◆

Activity 14.4

The first term of an arithmetic sequence is a and the third term is b, find the second term.

Homework Section 14-2

1. Find the third term of the arithmetic sequence 5, -2, ...

2. Find the 100th term of an arithmetic sequence 3, 4, 5, 6, ...

3. In an arithmetic sequence whose difference is d every term is multiplied by 3. Do you get an arithmetic sequence? If so, what is the difference?

4. If you take two arithmetic sequences and add corresponding terms, is the resulting sequence arithmetic? If so, what is the difference?

5. The first term of an arithmetic sequence is 4 and the difference is 7, what is the 60th term?

6. The first term of an arithmetic sequence is 3 and the fourth term is 21. What is the third term?

Definition: A *geometric sequence* is a sequence of numbers where each term is the product of the preceding one and a fixed number. This fixed number is called the *common ratio* of the geometric sequence.

Note: Sometimes the common ratio is also called the *growth factor*.

Example 14-3.1: The sequence 2, 6, 18, 54, 162, ... is a geometric sequence. The common ratio in this example is 3.

$\blacklozenge$

Activity 14.5

Let's write out the long form of the geometric sequence in the last example.

1st term	$2 = 2 \cdot 3^0$
2nd term	$6 = 2 \cdot 3^1$
3rd term	$18 = 2 \cdot 3^2$
4th term	$54 = 2 \cdot 3^3$

Looking for a pattern, what would be the 100th term of the sequence?

Find the 100th term of a geometric sequence whose first term is a_1 and whose common ratio is r.

We can express the definition of a geometric sequence in terms of the recursive formula

$$a_n = r a_{n-1}$$

where a_n is any term, a_{n-1} is the preceding term, and r is the common ratio.

If we know the first term, we can find any other term by multiplying by the common ratio a sufficient number of times. For example, writing the first five terms of a geometric sequence with first term a_1 and common ratio r, we get

$$a_1, \ a_1 r, \ a_1 r^2, \ a_1 r^3, \ a_1 r^4.$$

In general , the expression for the **nth term** of a *geometric sequence* is

$$a_n = a_1 r^{n-1}$$

Example 14-3.2: Find the common ratio and the number of terms in the
geometric sequence

$$6, \ 2, \ \frac{2}{3}, \ \frac{2}{9}, \ ..., \frac{2}{2187}$$

Solution: We know that in a geometric sequence to go from one term to the
next we multiply by the common ratio r. So to get the second term
we have to multiply 6 by r. This would give the equation

$$2 = 6r.$$

Solving for r, the common ratio, we get $r = \dfrac{1}{3}$. To find the number
of terms in the sequence we can use the previous formula,
$a_n = a_1 r^{n-1}$. In this example, $a_1 = 6$ and $r = \dfrac{1}{3}$ giving us the equation

$$\frac{2}{2187} = 6\left(\frac{1}{3}\right)^{n-1}$$

We can use Excel to solve this equation. You can see in the
following spreadsheet that there would be 9 terms in the
sequence.

	A	B	C	D	E
1	term	value of term			
2	1	6			
3	2	2		2/2187 =	0.000914495
4	3	0.666666667			
5	4	0.222222222			
6	5	0.074074074			
7	6	0.024691358			
8	7	0.008230453			
9	8	0.002743484			
10	9	0.000914495			
11	10	0.000304832			
12	11	0.000101611			

Activity 14.6

Find the third term of the geometric sequence

$$3, 4, \ldots$$

Activity 14.7

The first term of a geometric sequence is 1 and the third term is 4. What is the second term?

Example 14-3.3: The first term of a geometric sequence is a and the third term is b. Find the second term.

Solution: Assume that s is the second term. We know that if the common ratio is r the first three terms of the sequence are a, s, b. We can write s and b in terms of a.

$$s = ar \ \text{ and } \ b = (ar)r = ar^2$$
$$s^2 = (ar)^2 = a^2 r^2 = aar^2 = ab$$

Therefore, $s = \sqrt{ab} \ \text{ and } \ s = -\sqrt{ab}$. ♦

Homework Section 14-3

1. Find the fourth term of a geometric sequence where the first term is 4 and the second term is 8.

2. Find the common ratio of the geometric sequence

$$3.2,\ 4.8,\ 7.2,\ 10.8,\ \dots\ .$$

3. The first term of a geometric sequence is 14 and the recursive formula

$$a_n = 0.3a_{n-1}\ .$$

Write the general expression for the nth term of the sequence. What is a_{20}?

Suppose a friend told you that her first year salary was $42,000 and that each year she received annual raises averaging $2,000. What was the *total* income she received the first 15 years? A sum whose terms form an arithmetic sequence is called an *arithmetic series*. In this example, we are looking for the sum 42,000 + 44,000 + 46,000 + 48,000 + ... + 70,000. We can use Excel to find this sum by first entering the numbers 42,000, 44,000, ..., 70,000 and then entering =SUM(A1:A15) to get a sum of 840,000.

	A
1	42000
2	44000
3	46000
4	48000
5	50000
6	52000
7	54000
8	56000
9	58000
10	60000
11	62000
12	64000
13	66000
14	68000
15	70000
16	=SUM(A1:A15)

It would also be beneficial if we had an explicit formula to compute the sum directly. Let's look at the next two examples as we develop an explicit formula.

Example 14-4.1: Find the sum of the following arithmetic series

$$2 + 4 + 6 + 8 + 10.$$

Solution: There are only five terms in this series so it would be easy to just add the terms, but let's do it using a technique that we can generalize to any arithmetic series.

First we write the series forward and backwards where S is the sum.

$$
\begin{aligned}
S &= 2 + 4 + 6 + 8 + 10 \\
S &= 10 + 8 + 6 + 4 + 2 \\
\hline
2S &= 12 + 12 + 12 + 12 + 12
\end{aligned}
\quad \text{(The sum of each column is 12)}
$$

There are 5 columns so 2S = 5(12) = 60. Therefore S = 30. ♦

Example 14-4.2: Compute the sum $1 + 3 + 5 + 7 + ... + 999$.

Solution: Trying the technique in the last example we get

$$
\begin{array}{rcccccccc}
S = & 1 + & 3 + & 5 + & 7 + & ... & + 997 & + & 999 \\
S = & 999 + & 997 + & 995 + & 993 + & ... & + \ 3 & + & 1 \\
\hline
2S = & 1000 + & 1000 + & 1000 + & ... & + & 1000 & + & 1000
\end{array}
$$

In order to complete the problem we need to know how many terms there are in the sum. Recall from section 2 that the nth term of an arithmetic sequence given by the formula $1 + 2(n-1)$. Since we know the nth term is equal to 999, we solve the equation $1 + 2(n-1) = 999$ and get $n = 500$. Since the series has 500 terms, $2S = 1000(500)$ or $S = 250,000$.

$\blacklozenge$

To generalize the last two examples, suppose the first term of an arithmetic sequence containing n terms is a and the nth term is b. Also the common difference is d. Writing the series twice we get

$$
\begin{array}{rcccccccc}
S = & a & + & (a+d) & + & (a+2d) & + & ... & + & (b-2d) & + (b-d) & + & b \\
S = & b & + & (b-d) & + & (b-2d) & + & ... & + & (a+2d) & + (a+d) & + & a \\
\hline
2S = & (a+b) & + & (a+b) & + & (a+b) & + & ... & + & (a+b) & + (a+b) & + & (a+b)
\end{array}
$$

Each column sums to $(a + b)$ and there are n columns so

$$2S = n(a + b).$$

Therefore, $S = \dfrac{n(a+b)}{2}$.

The sum of an **arithmetic series** is
$$S = \frac{n(a+b)}{2},$$
where n is the number of terms, a is the first term, and b is the last term.

Looking back at the opening example, we can now find the sum using the above definition.

Example 14-4.3: Find the sum of the arithmetic series

$$42,000 + 44,000 + 46,000 + 48,000 + ... + 70,000.$$

Solution: Here a $= 42,000$, b $= 70,000$, and n $= 15$. So by the formula above

$$S = \frac{15(42,000 + 70,000)}{2} = \frac{15(112,000)}{2} = 840,000 \qquad \blacklozenge$$

Let's look at how we can find the sum of a geometric series. For example, find the sum of the geometric sequence

$$1, 2, 4, 8, 16, \ldots, 128.$$

We will do something similar to what we did with an arithmetic series.

$$S = 1 + 2 + 4 + 8 + \ldots + 128 \qquad \text{(Multiply both sides by the common ratio)}$$
$$2S = 2 + 4 + 8 + \ldots + 128 + 256$$

Now subtract the first equation from the second.

$$S = 256 - 1 = 255.$$

Let's use this process to find an explicit formula for a general geometric series.

Example 14-4.4: The first term of a geometric sequence is a and its common ratio is r. Find the sum of the first n terms of this sequence.

Solution: Let's denote the sum by S:

$$S = a + ar + \ldots + ar^{n-2} + ar^{n-1}$$

Multiply both sides by r.

$$rS = ar + ar^2 + \ldots + ar^{n-2} + ar^{n-1} + ar^n$$

Now subtract the first equation from the second.

$$rS - S = (ar + ar^2 + \ldots + ar^{n-2} + ar^{n-1} + ar^n) - (a + ar + \ldots + ar^{n-2} + ar^{n-1})$$

We are left with

$$rS - S = ar^n - a$$

$$(r - 1)S = a(r^n - 1) \qquad \text{(factoring both sides of the equation)}$$

$$S = \frac{a(r^n - 1)}{(r - 1)} \qquad \text{(dividing both sides by } (r - 1))$$

Note: When $r = 1$, S would be undefined. In this case all the terms would be equal and the sum would be na.

♦

The sum of a **geometric series** is

$$S = \frac{a(r^n - 1)}{(r-1)},$$

where a is the first term of the sequence, r is the common ratio, and n is the number of terms in the series.

Example 14-4.5: Find the sum of the following geometric series.

$$3 + (-6) + 12 + (-24) + \ldots + 3(-2)^{n-1}$$

for $n = 8$.

Solution: For this series $a = 3$ and $r = -2$.
Using the formula above we get

$$S = 3\left(\frac{(-2)^8 - 1}{-2 - 1}\right) = -255.$$ ◆

Activity 14.8

Find the sum of the following series.

a) $1 + 2 + 3 + 4 + \ldots + 100$

b) $1 + \dfrac{1}{2} + \dfrac{1}{4} + \dfrac{1}{8}$

Homework Section 14 - 4

1. Find the sum of the following series using Excel.

   a) $\qquad$ $3 + 8 + 13 + \ldots + 88$

   b) $\qquad$ $5 + \dfrac{5}{2} + \dfrac{5}{4} + \dfrac{5}{8}$

2. Find the sum of the following geometric series.

   $$3 + (-6) + 12 + (-24) + \ldots + 3(-2)^{n-1}$$

   for $n = 10$.

3. Find the sum of the arithmetic series

   $$16 + 20 + 24 + \ldots + 108.$$

Section 14-5: Sigma Notation

The symbol $\displaystyle\sum_{i=1}^{n} y_i$ (Σ is the Greek capital letter **sigma**) tells us to sum the elements that appear to the right of the Σ sign, commencing with $y_1 \, (i.e., i = 1)$ and proceeding in order to y_n.

Thus for example,

$$\sum_{i=1}^{3} y_i = y_1 + y_2 + y_3 .$$

Sigma notation was developed because writing out the terms of a sum can be time consuming.

Example 14-5.1: Write the following sum in sigma notation.

$2 + 4 + 8 + 16 + 32 + 64 + 128 + 256 + 512 + 1024 + 2048 + 4096$
(There are **12** terms in the sum and this is a geometric series.)

Solution: The sum can be written as $\displaystyle\sum_{i=1}^{12} 2 \cdot 2^{i-1}$. This is because every term in the sum is of the form $2 \cdot 2^{i-1}$, and all the values of $2 \cdot 2^{i-1}$, from $i = 1$ to $i = 12$, are added up. In this example, "$2 \cdot 2^{i-1}$" is called the **general term**; 12 and 1 are the top and bottom **limits** of the sum. Notice that if the limits go from 1 to 12, you will have 12 terms in your sum.

$\blacklozenge$

Example 14-5.2: Evaluate $\displaystyle\sum_{i=1}^{4} i^3$

Solution:

$$\sum_{i=1}^{4} i^3 = 1^3 + 2^3 + 3^3 + 4^3 \qquad \text{(4 terms, since the limits from 1 to 4)}$$

$$= 1 + 8 + 27 + 64$$

$$= 100 \qquad\qquad \blacklozenge$$

<u>**Example 14-5.3**</u>: Evaluate $\displaystyle\sum_{n=1}^{5}(2+2(n-1))$

Solution: Here the limits go from 1 to 5, so n takes on the values 1, 2, 3, 4, and 5. The sum will therefore have 5 terms.

$$\sum_{n=1}^{5}(2+2(n-1)) = (2 + 2(\mathbf{1} - 1)) + (2 + 2(\mathbf{2} - 1)) + (2 + 2(\mathbf{3} - 1)) + (2 + 2(\mathbf{4} - 1))$$

$$+ (2 + 2(\mathbf{5} - 1)) = 2 + 4 + 6 + 8 + 10 = 30$$

♦

Activity 14.9

Write these expressions using $\sum$ notation.

a) $10 + 11 + 12 + \ldots + 50$

b) $1 + \dfrac{1}{2} + \dfrac{1}{4} + \ldots + \dfrac{1}{128}$

Homework Section 14 - 5

1. Evaluate $\displaystyle\sum_{i=2}^{5} i^3$.

2. Evaluate $\displaystyle\sum_{n=1}^{4} (n-1)^2$.

3. Write these expressions using Σ notation.

   a) $4 + 6 + 8 + \ldots + 16$

   b) $3 + \dfrac{3}{2} + 1 + \ldots + \dfrac{1}{3}$

<u>***Section 14-6: Applications***</u>

Example 14-6.1: Suppose you bought a new car for $20,000 and it depreciates 20% each year. What will the car be worth in five years? Generalize the result to find an explicit formula for the value of a new car which cost v and depreciates r% each year.

Solution: First let's set up a table for the first part of the problem.

Nth year	Value ($)
1	20,000
2	$20{,}000 - 20{,}000(.2) = 20{,}000(1 - .2) = 20{,}000(.8) = 16{,}000$
3	$20{,}000(.8) - 20{,}000(.8)(.2) = 20{,}000(.8)(1 - .2) = 20{,}000(.8)^2 = 12{,}800$
4	$20{,}000(.8)^2 - 20{,}000(.8)^2(.2) = 20{,}000(.8)^2(1-.2) = 20{,}000(.8)^3 = 10{,}240$
5	$20{,}000(.8)^3 - 20{,}000(.8)^3(.2) = 20{,}000(.8)^3(1-.2) = 20{,}000(.8)^4 = 8{,}192$

So after five years the car would be worth $8,192.

Generalizing this result, an explicit formula for a new car costing v and depreciating r% each year would be

$$v_n = v_1\left(1 - \frac{r}{100}\right)^{n-1}$$

where v_n is the value of the car after the nth year, v_1 is the initial value of the car, and r% is the depreciation rate of the car each year.   ♦

Example 14-6.2: Use Excel to answer the following question. Suppose that two cows on a ranch are infected with a certain disease. Suppose each infected cow infects two other cows each day. If there are 10,000 cows on the ranch, how long will it take before they are all infected? (Kime and Clark, p351)

Solution: Here is an Excel worksheet showing the formulas in column B and the actual number of infected cows in column C. The printout shows that on the 9th day all the cows will be infected.

	A	B	C
1	Day	Number of Infected Cows	
2	1	2	2
3	2	=B2+2*B2	6
4	3	=B3+2*B3	18
5	4	=B4+2*B4	54
6	5	=B5+2*B5	162
7	6	=B6+2*B6	486
8	7	=B7+2*B7	1458
9	8	=B8+2*B8	4374
10	9	=B9+2*B9	13122

♦

As before, let's try and generalize the process in the last example.

<table>
<tr><td>

Activity 14.10

In the last example find an explicit formula for the number of cows infected after n days.

</td></tr>
</table>

Our previous discussion of compound interest was concerned with lump sum investments, where we initially invested a fixed amount and let it sit and collect interest for a number of years. Frequently situations involve a series of equal periodic payments rather than lump sums. These are known as annuities. An annuity is the payment or receipt of equal cash amounts per period for a specified amount of time. Examples of annuities are retirement plans, loan payments, and savings plans for future events like going to college.

Example 14-6.3: Suppose your grandfather deposits $3000 in an account for you each December 31 for the next four years and the money earns 4% per year. How much will you have at the end of four years?

Solution: Let's set up a table to keep track of the deposits.

End of year	Amount of Deposit	Number of Years Earning Interest	Future Value (Value at the End of 4 Years)
1	3,000	3	$3000(1.04)^3 = 3,374.59$
2	3,000	2	$3000(1.04)^2 = 3,244.80$
3	3,000	1	$3000(1.04)^1 = 3,120.00$
4	3,000	0	$3000(1.04)^0 = 3,000.00$

To get the amount after 4 years we have to add the column on the right. Writing the terms out we get

$$3000(1.04)^3 + 3000(1.04)^2 + 3000(1.04)^1 + 3000 = \$12,739.39.$$

This sum is a geometric series with $a = \$3,000$, $r = 1.04$, and $n = 4$. The sum, which in this case is the total future value, is the sum of the series, $\dfrac{a(r^n - 1)}{(r - 1)}$. The future value is then

$$\frac{\$3,000(1.04^4 - 1)}{(1.04 - 1)} = \$12,739.39.$$

♦

Activity 14.11

Suppose your grandfather deposited the money on January 1 of each year instead of December 31. How much would be in the account at the end of 4 years, assuming no withdrawals?

Note: When payment is at the beginning of each interest period the annuity is called an *annuity due*, when payments are at the end of a period it is called an *ordinary annuity*.

Unit 14 Review Exercises

1. Determine whether the sequence is arithmetic. If it is arithmetic, find the common difference.

    a) 3, 6, 9, 13, ...   b) 2, 4, 6, 8, ...   c) $3, \dfrac{3}{2}, 0, -\dfrac{3}{2}, ...$

2. Determine the common difference, the sixth term, the nth term, and the 100th term of the arithmetic sequence.

    a) 1, 5, 9, 13, ...   b) -12, -8, -4, 0, ...

    c) 2, 2 + s, 2 + 2s, 2 + 3s, ...

3. Which term of the arithmetic sequence 1, 4, 7, ... is 94?

4. The 100th term of an arithmetic sequence is 299, and the common difference is 3. Find the first three terms.

5. Use Excel to evaluate the following:

    a) $\displaystyle\sum_{n=0}^{5}(1-2n)$ b) $\displaystyle\sum_{n=0}^{20}(3+0.25n)$

6. Find the sum of the arithmetic series 1 + 5 + 9 + ... + 401.

7. Suppose you get a job with a salary of $41,000 a year. In addition, you are promised a $2,500 raise each year. Find your total earnings for a 10-year period.

8. Determine whether the sequence is geometric. If it is geometric, find the common ratio.

    a) $3, \dfrac{3}{2}, \dfrac{3}{4}, \dfrac{3}{8}, ...$ b) 27, -9, 3, -1, ...

9. Determine the common ratio, the fifth term, and the nth term of the geometric sequence.

    a) 2, 6, 18, 54, ...   b) $7, \dfrac{14}{3}, \dfrac{28}{9}, \dfrac{56}{27}, ...$

10. Find the 8th term of the geometric sequence 4, 12, 36, ...

11. The third term of a geometric sequence is $\dfrac{45}{4}$, and the fifth term is $\dfrac{405}{16}$. Find the fourth term.

12. Find the following sums.

   a) $1 + 3 + 9 + \ldots + 2187$

   b) $1 - \dfrac{1}{2} + \dfrac{1}{4} - \dfrac{1}{8} + \ldots - \dfrac{1}{512}$

   c) $\displaystyle\sum_{j=0}^{10} 3\left(\dfrac{1}{2}\right)^{j}$

13. A culture initially has 1000 bacteria, and its size increases by 5% every hour. How many bacteria are present at the end of 5 hours? How many are present at the end of n hours?

14. When a drug is introduced into the blood stream, the body has mechanisms to eliminate it. For example, some drugs are removed by the kidneys. How does the amount of the drug in the blood diminish over time? A commonly used model to describe this process starts with the assumption that a fixed percentage of the drug is removed every so many hours. For example, in the case of aspirin, about half is removed from the blood every half hour. Suppose you take two aspirins, 750 milligrams. Let a_n be the amount of aspirin in your blood after n half hour periods. Then $a_n = 0.5a_{n-1}$; $a_1 = 750$. How much aspirin remains in the blood after 4 hours. Find a general equation that doesn't need the previous term to find the amount after n half hour periods.

15. What can you say about a geometric sequence that has a common ratio or growth factor less than 1?

X Y
Z

UNIT 15
Vectors

A vector quantity is a quantity which is fully described by both magnitude and direction. On the other hand, a scalar quantity is a quantity which is fully described by just its magnitude. Scalars can be specified by single numbers (with the appropriate units) whereas vectors are quantities that cannot be described by single numbers. The temperature of the classroom is an example of a scalar quantity because we could say for instance, "the temperature is $73°$ F." When we express the distance an airplane has traveled as 25 miles north, we are giving a vector description because we have specified both the magnitude and the direction. This vector quantity is called displacement, which is the straight-line distance between two points, together with the direction. Other examples of vector quantities include velocity, acceleration and force. To fully describe each of these quantities both a magnitude and direction have to be specified.

Example 15-1.1: A scalar has only a numerical value and a unit; examples
are 5 inches, \$12, or 175 pounds. A vector has a
numerical value (with units) and a direction, such as
65 mph due north.   ◆

To determine whether a physical quantity is a vector we have to ask, "_Is the direction needed to fully describe this quantity?_" If the answer is yes, the quantity is a vector and if the answer is no, the quantity is a scalar. Vectors and scalars are manipulated differently. For example, in adding scalars all we use is ordinary addition. If you walked 15 meters East and then turned $90°$ and walked 15 meters North the total distance you traveled was 30 meters (15 + 15 = 30). But what is your total displacement? In other words, how far are you from your starting point? In terms of displacement, $15 + 15 \neq 30$. Your final displacement is 21.2 meters Northeast and is illustrated in the next example. To describe displacement, you must provide both the magnitude of the distance traveled and the direction.

Example 15-1.2: Suppose you walked 15 meters East and then 15
meters North, what is your displacement?

Solution: The displacement is the straight-line distance between the
starting point and the finishing point.

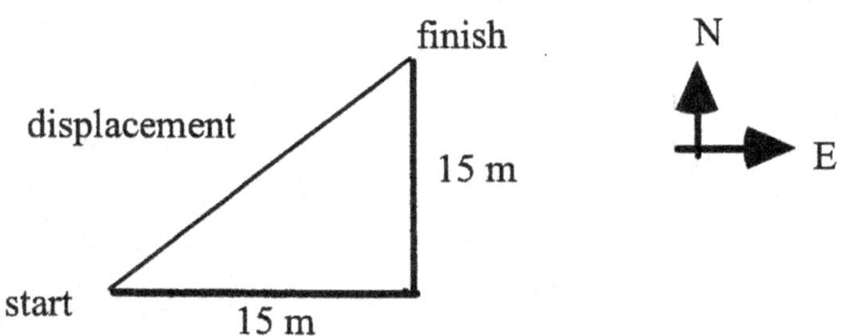

Since we have a right triangle we can use the Pythagorean Theorem to find the magnitude of the displacement.

magnitude of displacement = $\sqrt{15^2 + 15^2} = 21.2$ meters and the direction is to the Northeast. ♦

It is important to keep in mind the distinction between distance and displacement, distance is a scalar quantity and displacement is a vector quantity.

Activity 15.1

Suppose you walked 30 steps East, turned around and then walked 30 steps West.

a)   What is the distance you traveled?

b)   What is your total displacement?

You can see from the last example there is ordinary addition involving scalars, and vector addition, involving vectors. In ordinary addition, 30 + 30 is always 60, but in vector addition 30 + 30 depends on circumstances such as direction.

Section 15-2: Representation of Vectors

Vectors are represented in diagrams by arrows, as shown below. The vector magnitude is proportional to the length of the arrow and arrow direction indicates the vector direction.

LENGTH INDICATES
MAGNITUDE

ARROWHEAD INDICATES
DIRECTION

In addition, we will make a distinction between the two ends of the vector as indicated below.

Note: The position of a vector in space is unimportant. Two vectors are
equivalent if they have the same magnitude and direction.

Example 15-2.1: Below is an example of a properly drawn vector diagram.

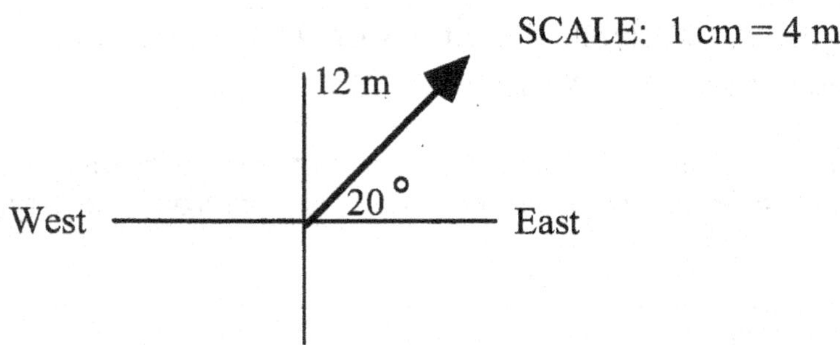

Some of the characteristics of a properly drawn vector diagram are:

- a scale is listed (*Here 1 cm = 4 m)*
- an arrow is drawn in a specified direction (*The direction is 20° to the north of east.*)
- the magnitude and direction of the vector is labeled. (*The magnitude of the vector is 12 m.*) ♦

The direction of a vector is often expressed as an angle of rotation about its starting point from east, west, north, or south. For example, a vector that is said to have a direction 20° North of East means the vector pointed East has been rotated 20° towards the North.

Another important vector is force. A force is a push or a pull that is exerted by one object on another. The English unit for force is the pound, "lb" and the metric unit is the Newton, "N." The effect of a force depends on its strength and its direction in relation to the force's point of application. Consequently, to describe a force, we have to give its magnitude (strength) and its direction.

Activity 15.2

Using a protractor and a ruler, draw a vector representing a force of 30-lb at 60° North of East and a second vector representing a force of 40-lb directly East. Select a scale, such as 1 cm = 10 lb.

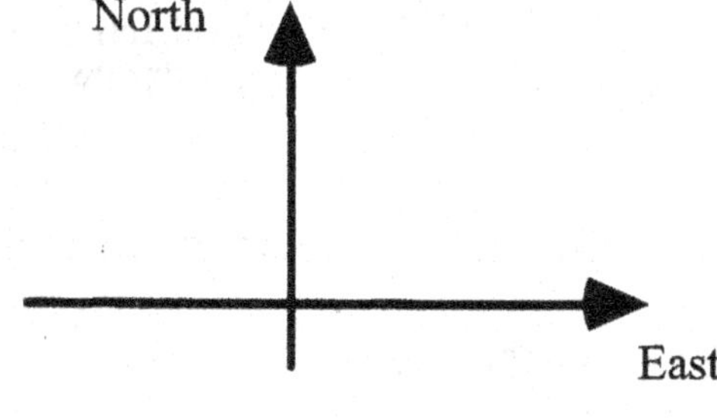

Activity 15.3

Give the magnitude and direction of the following vector, V. Assume vector V measures 3.8 cm.

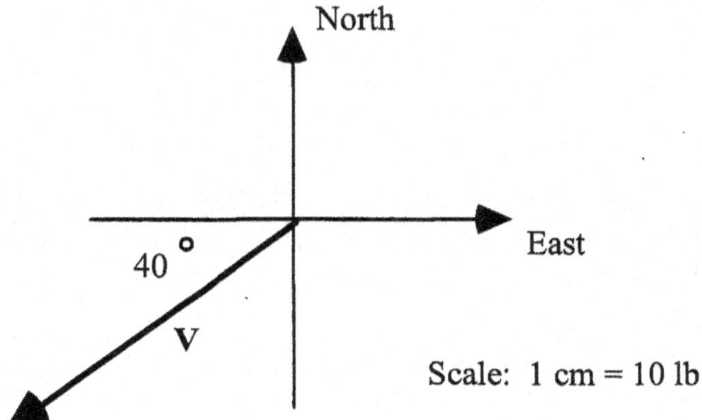

Here vector V represents a force.

Section 15-3: Vector Addition

We can perform a variety of operations with vectors. One such operation is the addition of vectors. Since force is a vector, two forces are added together by vector addition.

Example 15-3.1: Suppose two people are pushing on a cart, one with a 20 lb force and the other with a 28 lb force. What's the sum of their forces?

Solution: We are looking for a vector that will accomplish the same result as the two vectors. This single vector is called the resultant. We cannot find the resultant unless we know what direction the two people are pushing. If the two people were pushing in the same direction, the sum would be 48 lb as illustrated below.

Suppose the two people were pushing against each other. This can be represented by the following vector diagram. The resultant is 8 lb to the left.

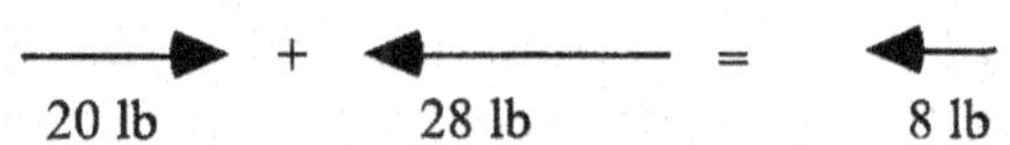

Activity 15.4

a)   Find the resultant of the following vector additions.

b)   Suppose a 50 lb force is applied to a body in one direction and exactly the same force is applied in the opposite direction, what is the sum of the two forces?

c)   Suppose two 50 lb forces are applied to a body in the same direction, what is the sum of the two forces?

d)   In vector addition, 50 lb + 50 lb can take on what range of values?

You can see from the last examples that if vectors are going in the same direction or the opposite direction it is easy to find the resultant. We want to be able to find the resultant in more complicated cases like the following.

To add these two vectors we will use the start-to-finish method. This is done by moving one vector, keeping it in the same direction, so that it begins where the first vector ends. Recall that vectors are equal as long as they have the same direction and magnitude, which allows us to move vectors without changing them. Once the vectors are drawn start-to-finish, the resultant is drawn from the start of the first vector to the finish of the last vector. This is illustrated in the following figure.

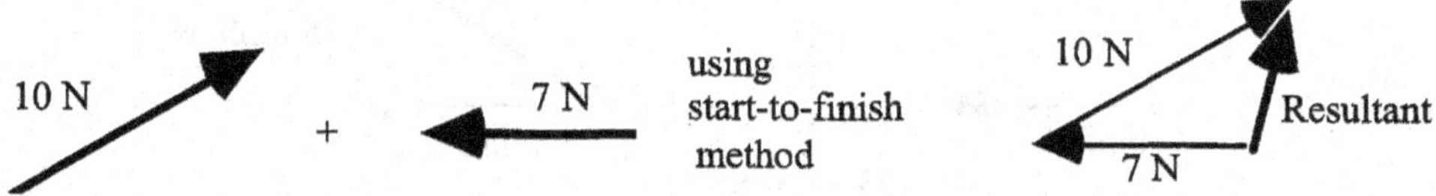

So when we add the two vectors above we get

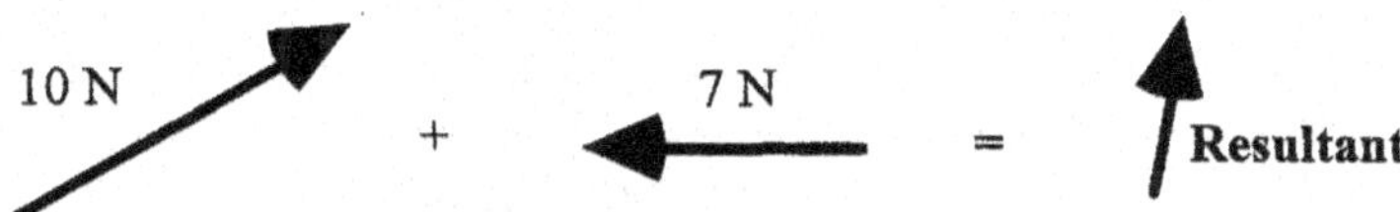

Let's continue by looking at adding vectors that are at right angles to each other.

Example 15-3.2: Use the start-to-finish method to add the two vectors **A** and **B**, where **A** is a vector with magnitude 50 mph East and **B** is a vector with magnitude 35 mph North.

Solution: The vectors here are not drawn to scale; we are merely illustrating the start-to-finish method.

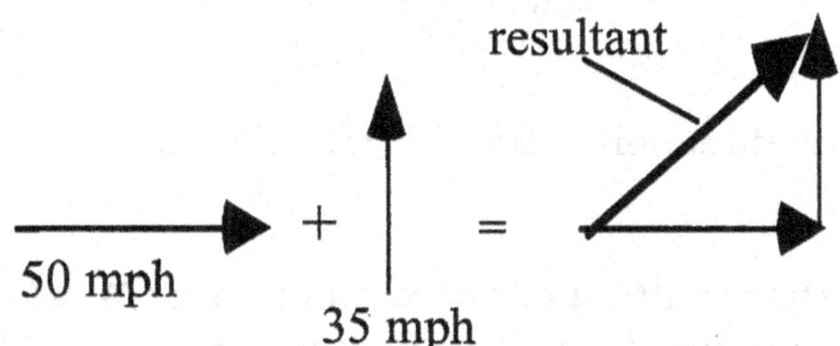

If these vectors were drawn to scale, we could use a ruler to measure the length of the resultant and determine its magnitude by converting to real units using the scale. The direction could be determined by using a protractor to measure the angle between the first vector and the resultant. The problem with this method is that it can be very inaccurate.

Instead of using a ruler and protractor to add the two vectors, we can use the Pythagorean Theorem to determine the resultant, which is the hypotenuse of the right triangle.

Now, let's use the Pythagorean Theorem to solve the previous example.

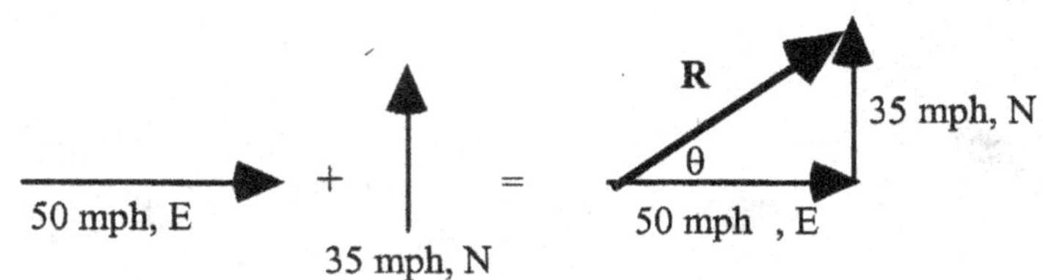

$$R^2 = 50^2 + 35^2$$

$$R^2 = 3725$$

$$|R| = 61 \text{ mph}$$

Note : The symbol $|R|$ stands for the magnitude of R.

The resultant, **R**, is a vector with a magnitude of 61 mph. To find the direction, we first need to find the angle made between the horizontal vector and the resultant. We have labeled that angle θ (theta). Since $\tan\theta = \dfrac{35}{50}$, we can determine θ by calculating

$$\theta = \tan^{-1}\left(\frac{35}{50}\right) = 35.0°.$$

Therefore, the direction is 35.0° North of East.

♦

Example 15-3.3: Find the resultant of two vectors **A** and **B**, where **A** is a vector of magnitude 7 N, East and **B** is a vector of magnitude 12 N, North. You can think of these vectors as forces pulling on an object.

Solution: First let's draw a diagram that represents the problem.

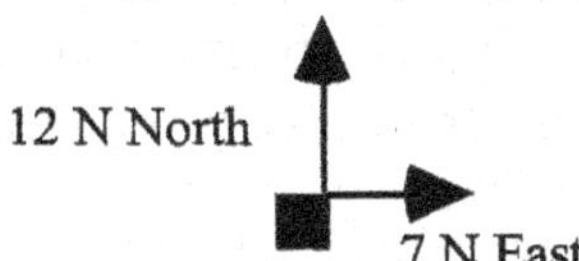

Now let's move **B** so that it starts where **A** finishes.

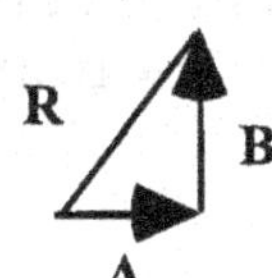

Since the vectors are perpendicular, we can use the Pythagorean Theorem to find **R**, the resultant vector.

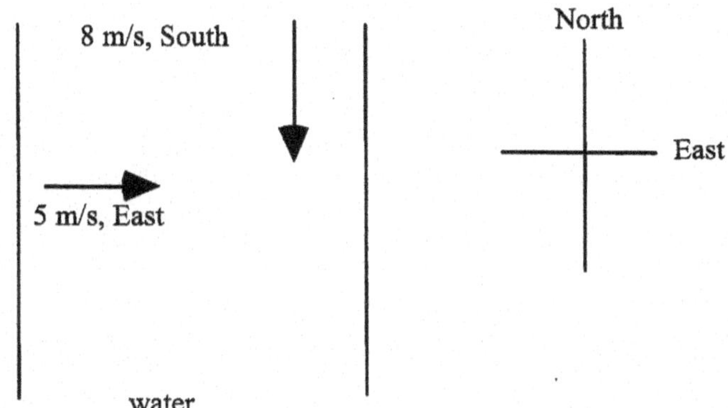

$$\mathbf{R}^2 = 7^2 + 12^2$$
$$\mathbf{R}^2 = 49 + 144 = 193$$
$$\mathbf{R} = \sqrt{193} = 13.9$$

The resultant has a magnitude of **13.9 N.** Now we have to find the direction. To find the direction we have to find θ in the figure above. From the figure we have that $\tan \theta = \dfrac{12}{7}$. So

$\theta = \tan^{-1}(\dfrac{12}{7}) = 59.7°$. The direction of **R** is therefore 59.7° **North of East.**　　　　♦

Example 15-3.4: A boat traveling 5 m/s, East encounters a current traveling 8 m/s, South. What is the resultant velocity of the boat?

Solution:　First let's draw a diagram.

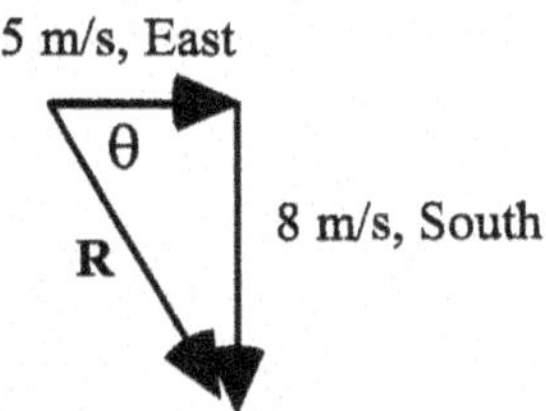

To add the vectors we will use the start-to-finish method.

Using the Pythagorean Theorem,

$$\text{The magnitude of } \mathbf{R} = \sqrt{5^2 + 8^2} = \sqrt{89} = 9.4 \text{ m/s}$$

To get the direction, $\theta = \tan^{-1}\left(\dfrac{8}{5}\right) = 58°$. Hence the direction is

58° South of East.　　　　♦

The figure below shows two vectors, one 12 lb, East and the other 5 lb, South. Find the resultant vector.

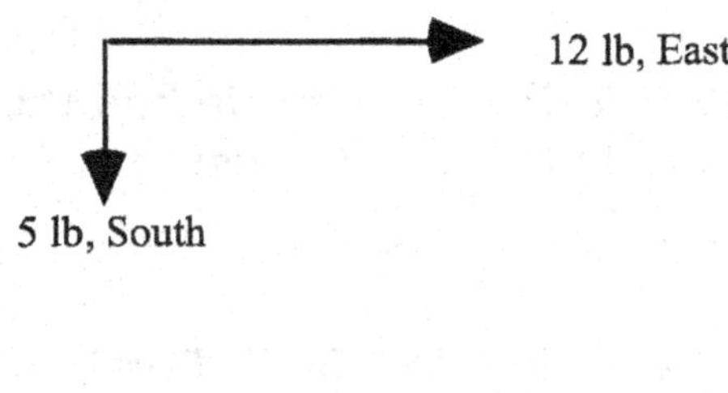

Homework Sections 15-1 - 15-3

1. Suppose you walked 30 steps East and then walked 20 steps South.

   a) What is the distance you traveled?

   b) What is your total displacement?

2. Give the magnitude and direction of the following vector, **V**. Assume vector **V** measures 1.5 cm.

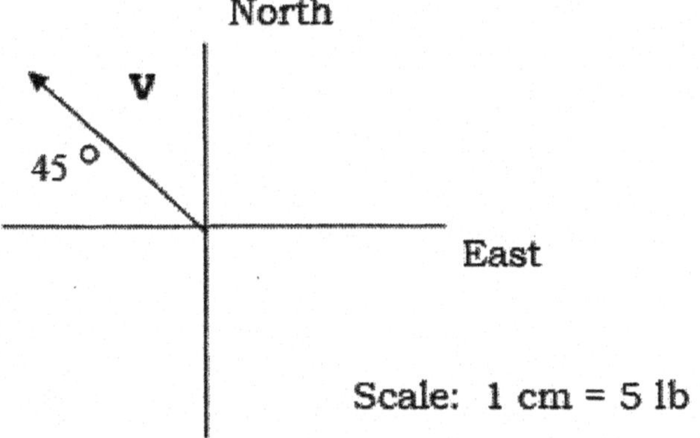

   Here vector **V** represents a force.

3. The figure below shows two vectors, one 10 lb, West and the other 5 lb, North. Find the resultant vector.

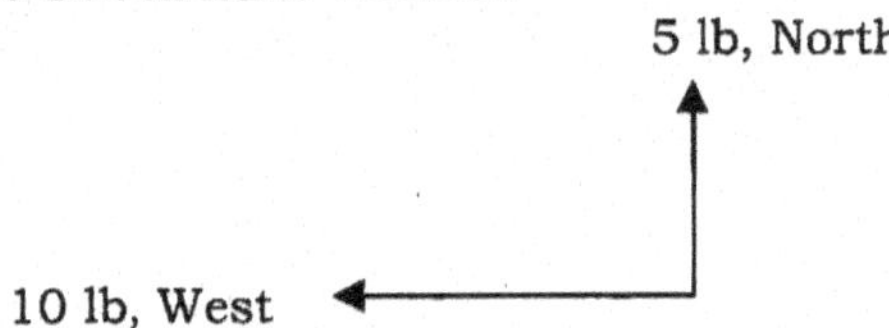

4. An airplane traveling 210 mph, East encounters a wind traveling 70 mph, South. What is the resultant velocity of the airplane?

We now have to learn how to add vectors that are not in the same direction, in the opposite direction, or perpendicular to each other. Before doing so, we have to introduce two special vectors, **i** and **j**. These vectors are both **unit vectors**, which means they have a magnitude of 1. They are illustrated below, and you can see that **i** is in the positive x direction and **j** is in the positive y direction.

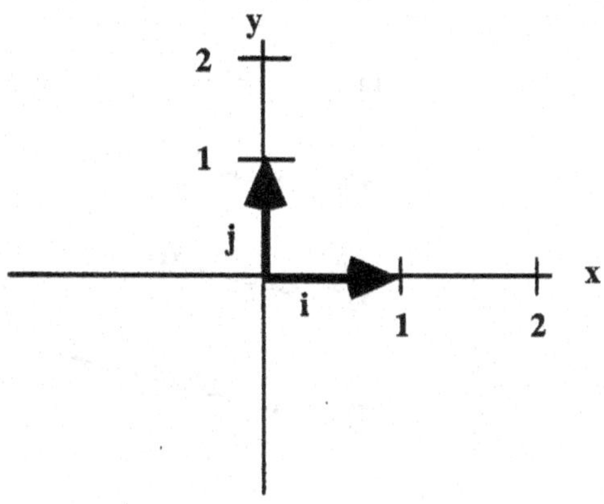

If a vector **B** is in the same direction as vector **A**, and **B** has magnitude n times that of **A**, then **B** = n**A**, where n**A** is called the **scalar multiple** of vector **A**.

Example 15-4.1: In the figure below 2**A** is a vector that is twice as long as **A**, but is in the same direction. -2**A** is a vector twice as long as **A**, but is in the opposite direction.

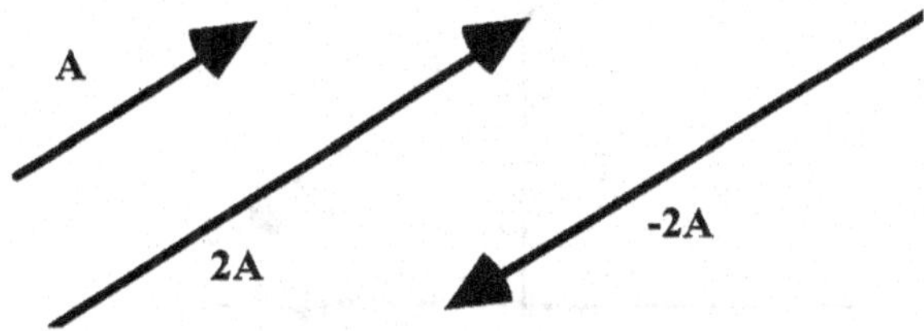

Referring back to the unit vector, if we have a vector 3**i**, then it is a vector of magnitude 3 that is in the positive x direction. Likewise 2**j** is a vector of magnitude 2 that is in the positive y direction.

Given vector **A** and **i** shown below, match the vector on the left with the vector diagram on the right.

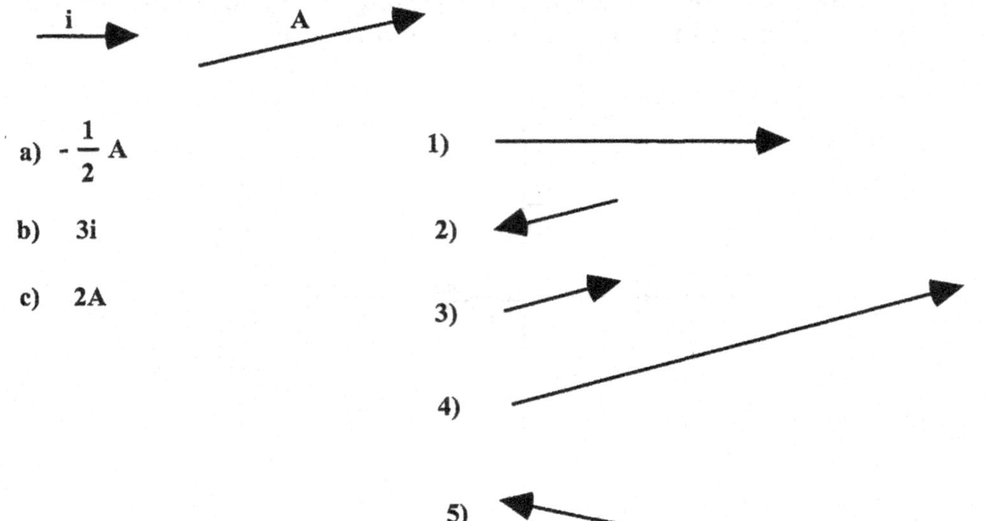

a) $-\dfrac{1}{2}$ **A**

b) 3**i**

c) 2**A**

1)

2)

3)

4)

5)

We want to be able to write a given vector **V** in terms of vectors **i** and **j**. These are called the **x- and y-components** of **V**. Given the vector **V** below, we can write it in terms of x- and y-components as **V** = 3**i** + 2**j**.

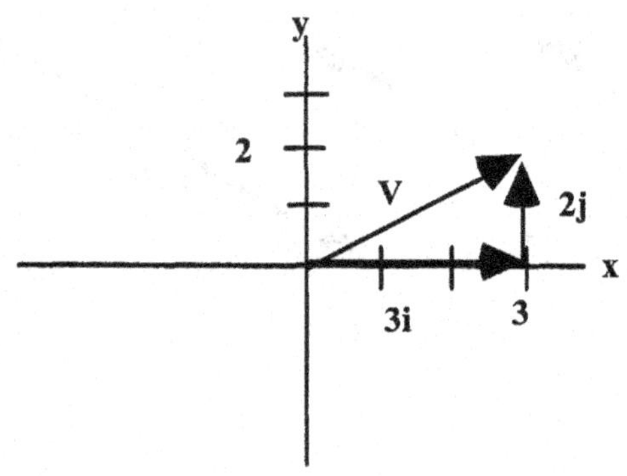

Example 15-4.2: Write the vectors **A**, **B** and **C** in x- and y-component form.

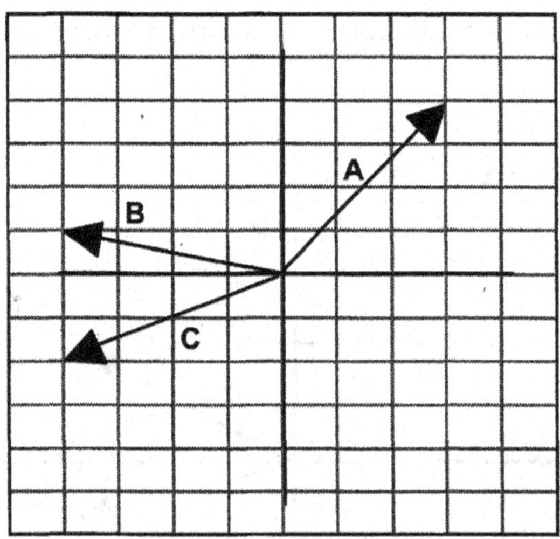

Solution: One way to write a vector in x-and y-component form is to start the vector at the origin and then look at the finishing point of the vector. In the case of vector **A**, the finishing point is (3, 4), so **A** = 3**i** + 4**j**.

The finishing point of **B** is (-4, 1), so **B** = -4**i** + **j**. Remember -4**i** is a vector of length 4 that is going in the negative x direction.

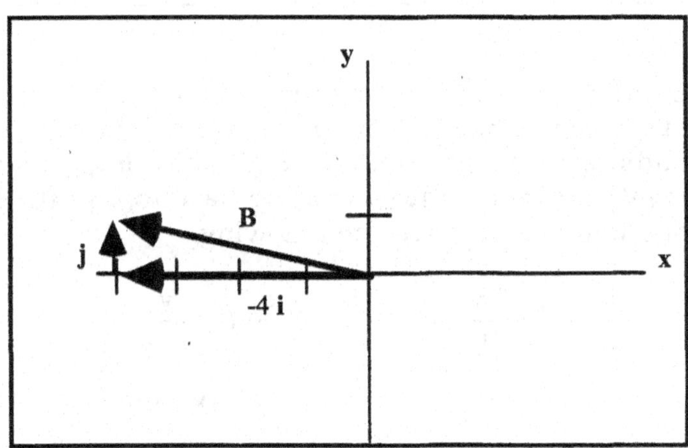

The finishing point of **C** is (-4, -2), so **C** = -4**i** -2**j**.

Example 15-4.3: Find the magnitude of vector **B** in the last example.

Solution: Remember the symbol for the magnitude of **B** is |**B**|. By the Pythagorean Theorem, |**B**| = $\sqrt{4^2 + 1^2} = \sqrt{16 + 1} = \sqrt{17} = 4.1$.

---

a)   Write vectors **A** and **B** below in x-and y-component form.

b)   Find $|\mathbf{A}|$ and $|\mathbf{B}|$.

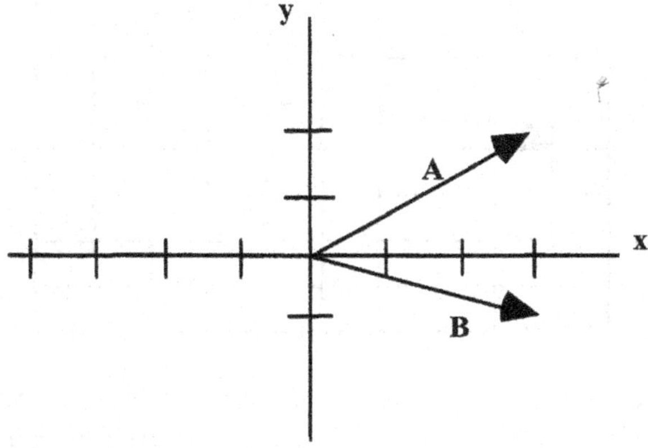

In most problems it is not easy to find the coordinates of the finishing point. Before we look at an example, let's review some trigonometry. We will have to find the length of the side opposite and adjacent to a given acute angle knowing the hypotenuse. In the figure below, we know $|\mathbf{V}|$ and the acute angle, we have to find the lengths x and y. From right triangle trigonometry we have the following.

$$\cos\theta = \frac{x}{|\mathbf{V}|} \qquad\qquad \sin\theta = \frac{y}{|\mathbf{V}|}$$

$$x = |\mathbf{V}|\cos\theta \qquad\qquad y = |\mathbf{V}|\sin\theta$$

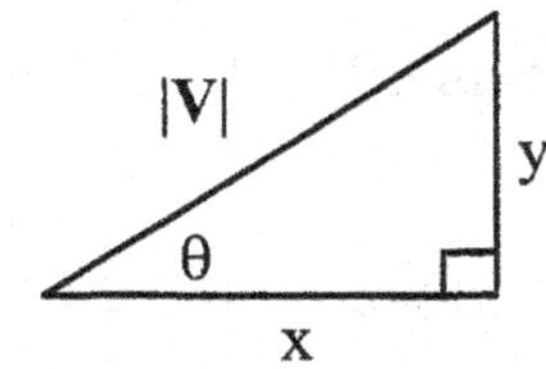

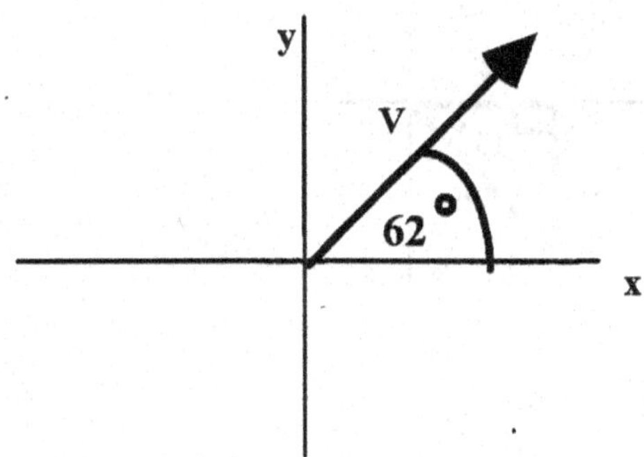

Solution: Let V_x represent the magnitude of the x-component and let V_y represent the magnitude of the y-component.

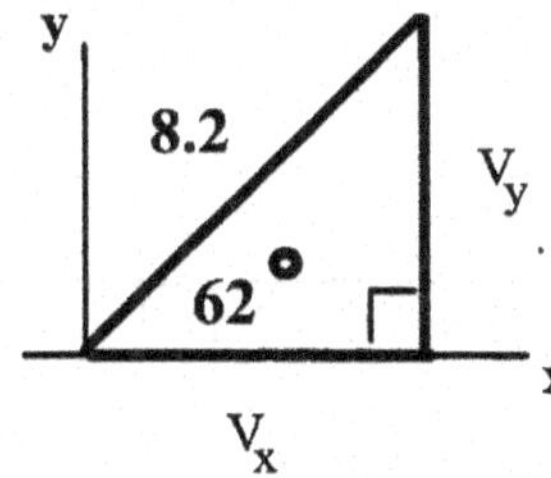

Using right triangle trigonometry, we have

$$\cos 62 = \frac{V_x}{8.2} \qquad \sin 62 = \frac{V_y}{8.2}$$

$$V_x = 8.2\cos 62 = 3.8 \qquad V_y = 8.2\sin 62 = 7.2$$

So, **V** = 3.8**i** + 7.2**j**.   ♦

Example 15-4.5: Suppose **A** is a vector of magnitude 20 and with direction $15°$ South of West. Find the x- and y- components of **A**.

Solution: First let's draw **A**.

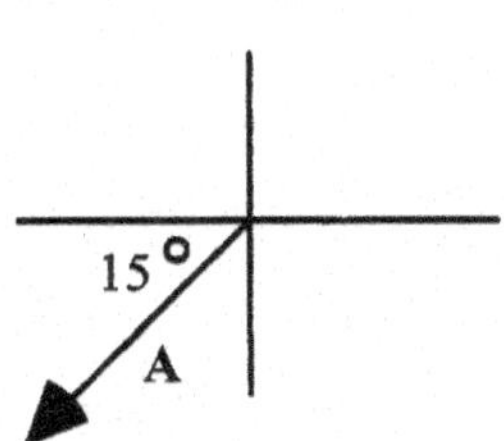

We are looking for A_x, the magnitude of the x- component and, A_y, the magnitude of the y- component.

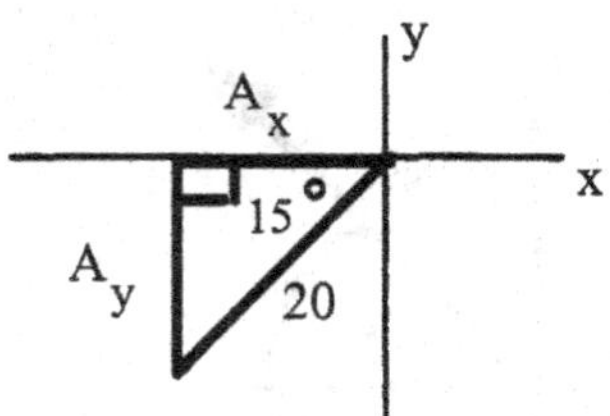

Solving for A_x and A_y, we get

$$\cos 15° = \frac{A_x}{20} \qquad \sin 15° = \frac{A_y}{20}$$

$$A_x = 20\cos 15° \qquad A_y = 20\sin 15°$$

$$A_x = 19.3 \qquad A_y = 5.2$$

Therefore, **A** = -19.3**i** - 5.2**j**.   ◆

Activity 15.8

Write vector **V** in component form, where $|\mathbf{V}| = 8.2$.

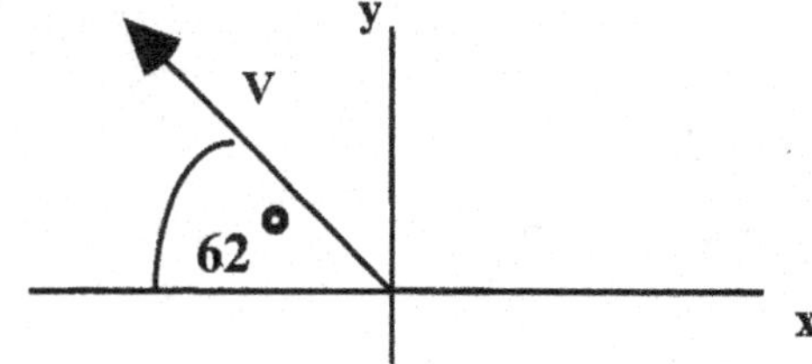

Homework Section 15-4

1. Write the vectors **A** and **B** in x- and y-component form.

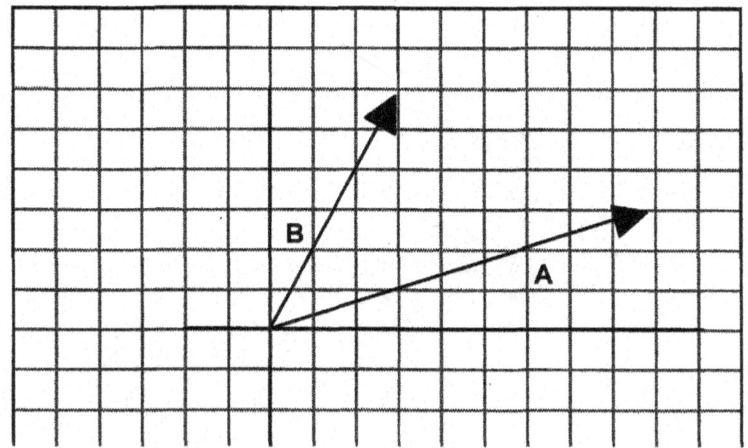

2. For the vectors in problem 1, find $|\mathbf{A}|$ and $|\mathbf{B}|$.

3. Write vector **V** in component form, where $|\mathbf{V}|$ = 5.1.

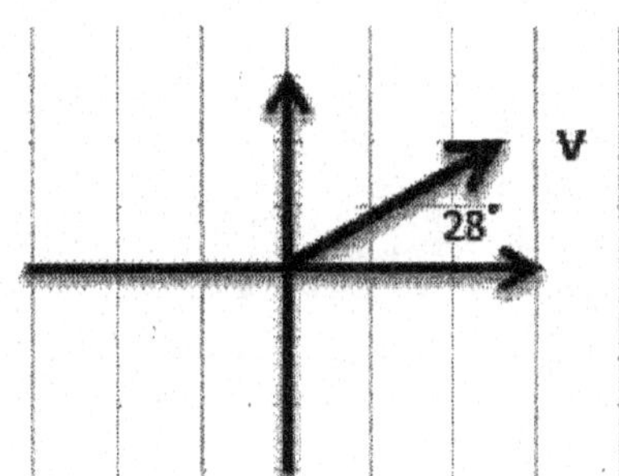

Suppose we had to add the following two vectors, **A** = 9**i** + 3**j** and **B** = 3**i** + 6**j**. These vectors are shown in the figure below.

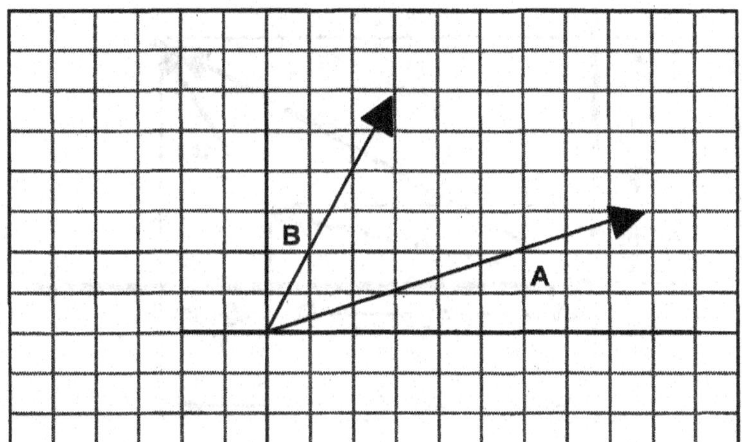

To add vectors **A** and **B**, we have to move one of the vectors, so the two vectors are start to finish. This is illustrated in the next diagram.

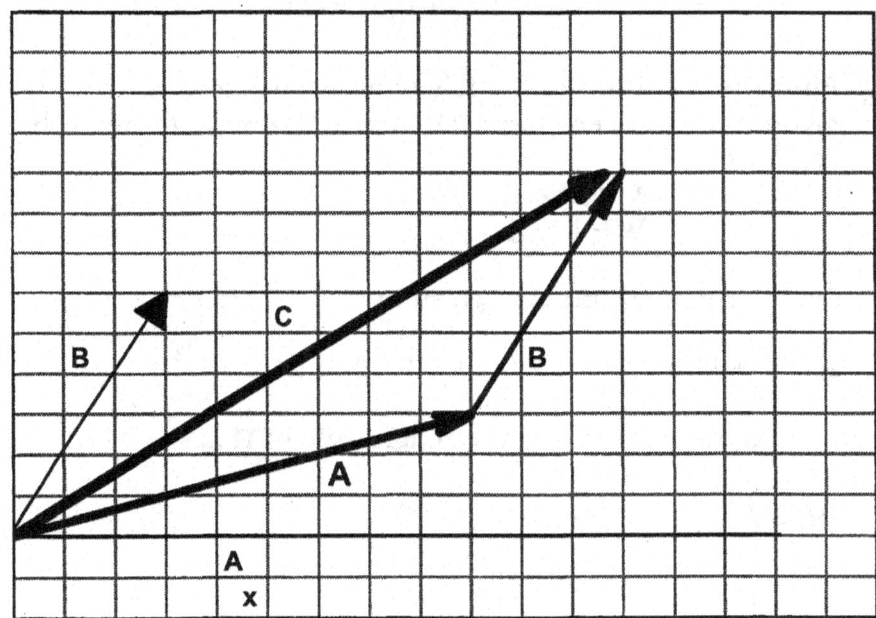

The x-component, or horizontal component of **A** was 9 and the horizontal component of **B** was 3. The horizontal component of **C**, the resultant, is 12, 9 + 3. We can get the vertical component, or y-component, of **C** in the same way; 3 + 6 = 9. So to get the resultant vector **C** in component form, we just have to add the x-components of **A** and **B** and the y-components of **A** and **B**; **C** = 12**i** + 9**j**. This is true when we have to add any vectors, write the vectors in component form and then add components. The next diagram illustrates this idea.

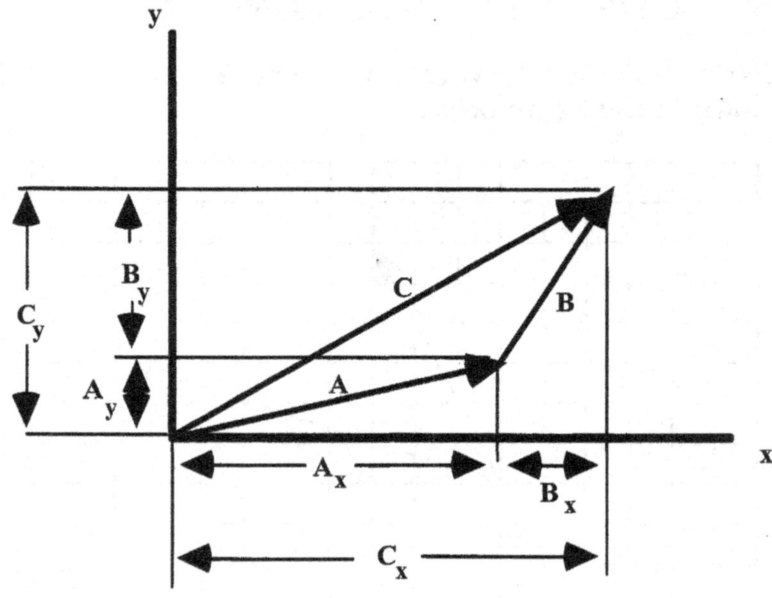

From the figure above, you can see that

$$C = (A_x + B_x)\mathbf{i} + (A_y + B_y)\mathbf{j} = C_x\mathbf{i} + C_y\mathbf{j}.$$

Example 15-5.1: Find the resultant of the following two vectors, which are given in component form. Then give the magnitude of the resultant.

$$\mathbf{V} = 2\mathbf{i} - 3\mathbf{j}$$
$$\mathbf{W} = \mathbf{i} + 5\mathbf{j}$$

Solution: From above, the resultant **R** is given by

$$\mathbf{R} = \mathbf{V} + \mathbf{W} = (2 + 1)\mathbf{i} + (-3 + 5)\mathbf{j} = 3\mathbf{i} + 2\mathbf{j}.$$

By the Pythagorean Theorem the magnitude of **R** is

$$|\mathbf{R}| = \sqrt{3^2 + 2^2} = \sqrt{13}.$$

Activity 15.9

Find the resultant of the following two vectors which are given in component form. Then give the magnitude of the resultant.

$$\mathbf{A} = 3.5\mathbf{i} - 6\mathbf{j} \text{ and } \mathbf{B} = -2\mathbf{i} + 9\mathbf{j}$$

Procedure for Adding Vectors by the Component Method

1. Resolve the vectors to be added into their vertical and horizontal components.

2. Add all the horizontal components together and all of the vertical components to obtain the horizontal and vertical components of the resultant.

3. Express the resultant using:
   a) the component form, or
   b) the magnitude-angle form.

Example 15-5.2: Find the resultant of the two forces below:

Force 1:   20 N   30° north of east
Force 2:   15 N   10° west of north

Solution: First draw a diagram representing the two vectors.

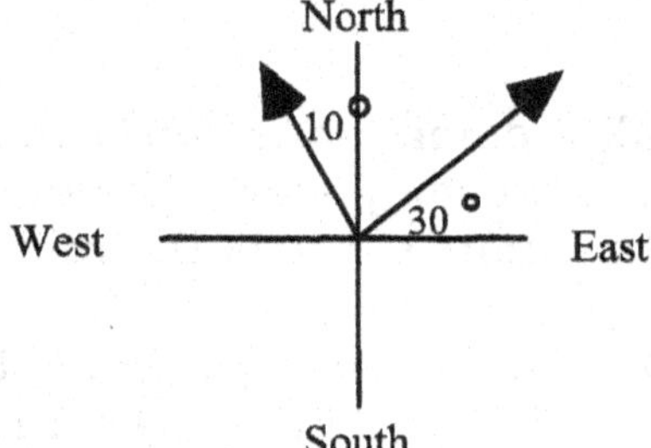

Step 1.   Resolve the vectors into their horizontal and vertical components.

Force 1: Call it vector **A**.

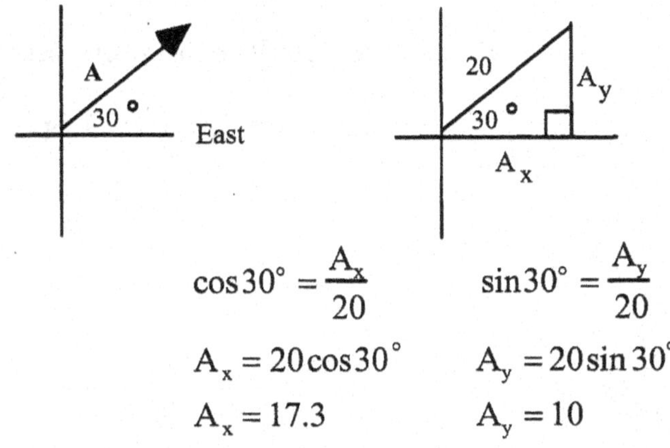

$$\cos 30° = \frac{A_x}{20} \qquad \sin 30° = \frac{A_y}{20}$$

$$A_x = 20\cos 30° \qquad A_y = 20\sin 30°$$

$$A_x = 17.3 \qquad A_y = 10$$

Therefore, **A** = 17.3**i** + 10**j**.

Force 2: Call it vector **B**.

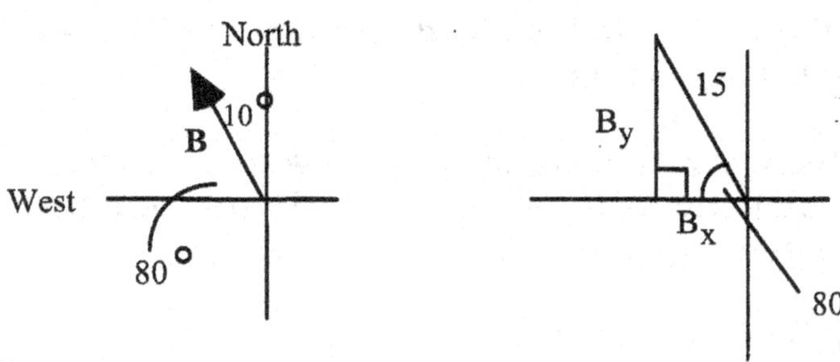

$$\cos 80^\circ = \frac{B_x}{15} \qquad \sin 80^\circ = \frac{B_y}{15}$$

$$B_x = 15\cos 80^\circ \qquad B_y = 15\sin 80^\circ$$

$$B_x = 2.6 \qquad B_y = 14.8$$

Therefore, **B** = -2.6**i** + 14.8**j**.

Step 2. Add all the horizontal components together and all of the vertical components to obtain the horizontal and vertical components of the resultant.

$$\begin{aligned}
\mathbf{A} + \mathbf{B} &= (17.3\mathbf{i} + 10\mathbf{j}) + (-2.6\mathbf{i} + 14.8\mathbf{j}) \\
&= (17.3 - 2.6)\mathbf{i} + (10 + 14.8)\mathbf{j} \\
&= 14.7\mathbf{i} + 24.8\mathbf{j}
\end{aligned}$$

Step 3. Express the resultant using:
      a) the component form, or
      b) the magnitude-angle form.

a) The resultant in component form is

$$\mathbf{R} = 14.7\mathbf{i} + 24.8\mathbf{j}.$$

b)   To express the resultant in magnitude-angle
form we first sketch the vector.

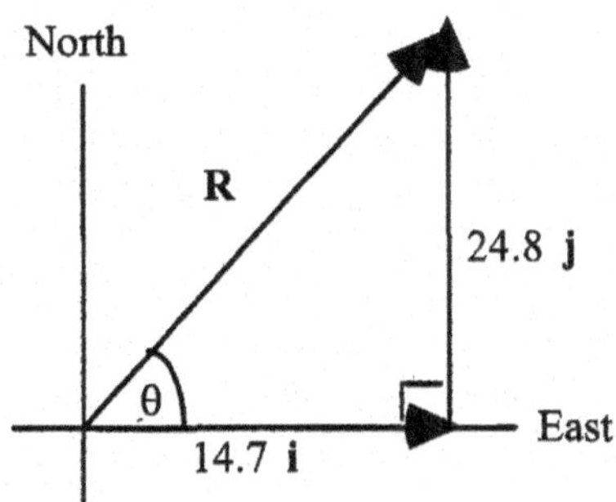

$$|\mathbf{R}| = \sqrt{14.7^2 + 24.8^2} = 28.8$$

$$\tan \theta = \frac{24.8}{14.7}$$

$$\theta = \tan^{-1}\left(\frac{24.8}{14.7}\right) = 59.3°$$

So the magnitude of R is 28.8 and the direction is
59.3° North of East.   ♦

Activity 15.10

A plane has a speed of 270 mi/hr in still air. It is flying in the direction 20° South of
East through a wind with a speed of 75 mi/hr in the direction 20° North of East. Find
the speed and direction of the plane.

Unit 15 Review Exercises

1.   Write the following vectors in component form

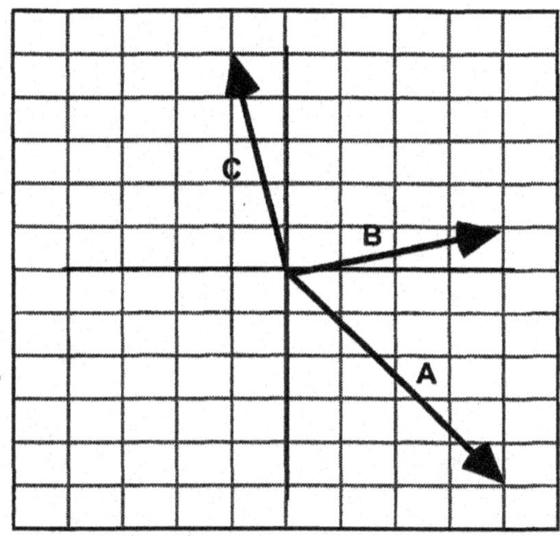

2.   Find the direction of vector **A** below.

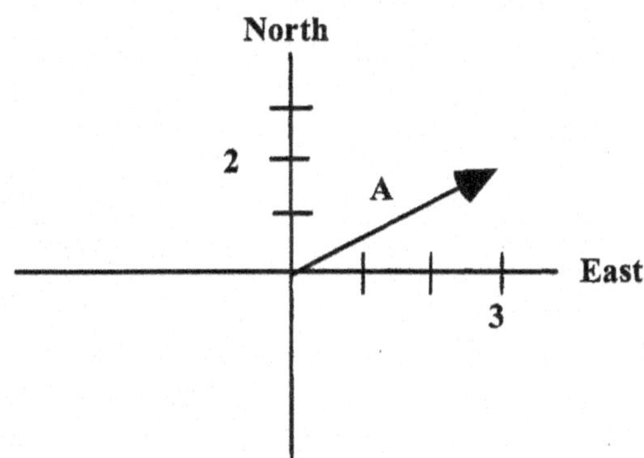

3.   Find the horizontal and vertical components of a vector **V** that has a magnitude of 12.0 m/s and is oriented at an angle of 60° relative to the positive x axis.

4.   Find the resultant of the two vectors **A** = 4.0 **i** + 2.0 **j** and **B** = -6.0 **i** + 3.5 **j**.

5.   Two forces, one 3 lb and the other 4 lb pull on an object. The angle between them is 90°. Find the magnitude and direction of the resultant vector.

Appendix
Solutions to the Activities

Activity 1.1

Let's look at the number of vertical and horizontal matches to determine if there is a pattern.

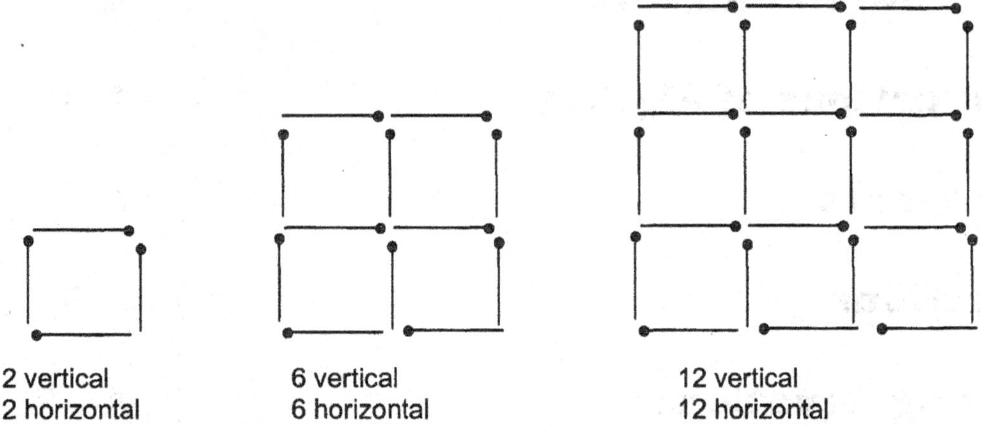

2 vertical
2 horizontal

6 vertical
6 horizontal

12 vertical
12 horizontal

Gives the sequence 4, 12, 24, 40, …

Letting $a_1 =$ the first term, $a_2 =$ the second term, $a_3 =$ third term, and $a_n =$ the nth term, then

$$a_1 = 2 \bullet 1 + 2 \bullet 1 = 4$$
$$a_2 = 3 \bullet 2 + 3 \bullet 2 = 12$$
$$a_3 = 4 \bullet 3 + 4 \bullet 3 = 24$$
$$a_4 = 5 \bullet 4 + 5 \bullet 4 = 20$$
$$a_n = (n+1)n + (n+1)n$$

Simplifying the above expression, we find that the number of matches needed to make N^2 unit squares would be $2n^2 + 2n$.

Activity 1.2

Let's suppose that today is June 1, 2012. Then d = 1, m = 6, and y = 2012. Substituting these values into the formula gives

$$W = 1 + 2(6) + \left\lfloor \frac{3(6+1)}{5} \right\rfloor + 2012 + \left\lfloor \frac{2012}{4} \right\rfloor - \left\lfloor \frac{2012}{100} \right\rfloor + \left\lfloor \frac{2012}{400} \right\rfloor + 2$$

$$= 1 + 12 + 4 + 2012 + 503 - 20 + 5 + 2 = 2519$$

2519/7 = 359 remainder 6. Since the remainder is 6, the day of the week is Friday.

Activity 2.1

   a) 453 + 16999 = 452 + (1 + 16999) = 17452   Associative Property of Addition

   b) 799 + 1457 + 101 = (799 + 1) + 100 + 1457 = 2357   Commutative Property of Addition and
                                                                   Associative Property of Addition

Activity 2.2

   a) $17 \bullet (25 \bullet 4) = 1700$ Associative Property of Multiplication

   b) $(125 \bullet 8) \bullet 37 = 37000$ Commutative Property of Multiplication

   c) $(1000 + 1) \bullet 20 = 20020$ Distributive Property of Multiplication over Addition

   d) $(1000 + 1) \bullet 102 = 102102$ Distributive Property of Multiplication over Addition

Activity 2.3

   a) $7p - 35 = 7(p - 5)$ b) $2\pi h - 4\pi = 2\pi(h - 2)$

Activity 2.4

a) $a^2 + 5ab + 6b^2 = a^2 + 2ab + 3ab + 6b^2 = a(a + 2b) + 3b(a + 2b) = (a + 3b)(a + 2b)$

b) $1 + a + a^2 + a^3 = 1(1 + a) + a^2(1 + a) = (1 + a^2)(1 + a)$

c) $1 + a + a^2 + a^3 + a^4 + a^5 = 1(1 + a) + a^2(1 + a) + a^4(1 + a) = (1 + a)(1 + a^2 + a^4)$

d) $1 + a + a^2 + a^3 + a^4 + a^5 + a^6 + a^7 = 1(1 + a) + a^2(1 + a) + a^4(1 + a) + a^6(1 + a)$
     $= (1 + a)(1 + a^2 + a^4 + a^6)$

e) pattern: $(1 + a)(1 + a^2 + a^4 + \ldots + a^{n-1})$ where n is an odd integer

Activity 2.5

a)

$$12 \div 6 \bullet 2 - 7(3 - -2)$$
$$2 \bullet 2 - 7(5)$$
$$4 - 35$$
$$-31$$

b)

$$4[6 - 4(2 - 7)$$
$$4[6 - 4(-5)]$$
$$4[26]$$
$$104$$

c)

$$5 \bullet 3^4 + 16 \div 8 - 2^2$$
$$5 \bullet 81 + 16 \div 8 - 4$$
$$405 + 2 - 4$$
$$407 - 4$$
$$403$$

d)

$$\frac{8 \bullet 2 - 4}{10 - 4 \bullet 3}$$
$$\frac{16 - 4}{10 - 12}$$
$$\frac{12}{-2}$$
$$-6$$

<u>Activity 2.6</u>

a) $\dfrac{49+14}{49+7+14} = (49+14)/(49+7+14)$

b) $\dfrac{\frac{3}{5}}{\frac{4}{7}} = (3/5)/(4/7)$

c) $\dfrac{f}{1-r} - s - f = f/(1-r) - s - f$

<u>Activity 2.7</u>

a) 11   b) 21   c) 196

<u>Activity 2.8</u>

a) $1.2*(4.5-8.4) = -4.68$

b) $(-3.2)*(4.6+5.9) = -33.6$

c) $(123*40.2)/(24.65-12.93) = 421.894$

<u>Activity 2.9</u>

a) 0.380952381   b) $1.1\overline{6}$ *or*   1.167

<u>Activity 2.10</u>

a) $\dfrac{2}{x} + \dfrac{3}{y} = \dfrac{2y+3x}{xy}$

b) $\dfrac{1}{x} - \dfrac{2}{x-3} = \dfrac{1(x-3)-2x}{x(x-3)} = \dfrac{x-3-2x}{x(x-3)} = \dfrac{-x-3}{x(x-3)}$

<u>Activity 2.11</u>

When $x = \dfrac{2}{3}$, $x(4x-30) = -18.\overline{2}$

<u>Activity 2.12</u>

$$A = \pi(2.5)^2 = 19.635\ cm^2$$

<u>Activity 2.13</u>

$$R = \frac{R_1 R_2}{R_1 + R_2}\quad \text{in Excel } (R_1 * R_2)/(R_1 + R_2) = 2500\ \text{ohms}$$

<u>Activity 2.14</u>

1. 1 year: $\$15305 - 0.27(\$15305) = \$11172.65$
   2 years: $\$11172.65 - 0.13(\$11172.65) = \$9720.2055 = \$9720.21$
   3 years: $\$9720.21 - 0.13(\$9720.21) = \$8456.5827 = \$8456.58$
   4 years: $\$8456.58 - 0.14(\$8456.58) = \$7272.6588 = \$7272.66$
   5 years: $\$7272.66 - 0.13(\$7272.66) = \$6254.4876 = \$6254.49$

   a) $11172.65
   b) $6254.49

2. $V = P(1-d_1)(1-d_2)(1-d_3)$
   After 3 years:
   $V = 15305(1-0.27)(1-0.13)(1-0.13)$
   $V = \$8456.58$

3. Multiply $15305 by 73%.
   $15305 \bullet (0.73) = \$11172.65$

Activity 3.1

expression	base	exponent
a) $(-5)^2$	-5	2
b) -5^2	5	2
c) $(x+y)^2$	x + y	2
d) $4x^3$	x	3

Activity 3.2

a) The base is $\{3 - 3(12 \div 6 \bullet 3 - 5 + 6)\}$. b) $-x^3$ is the same as choice i, ii and iii

Activity 3.3

Recall that in Excel you must start each expression with an equal sign to calculate.

a) $3 + 5\char94 2 = 28$

b) $(3+5)\char94 2 = 64$

c) $-(6\char94 2) - (3 - 4*6) = -15$

d) $((2 - 5\char94 2)/(2*5))\char94 2 = 5.29$

e) $5*(-18)\char94 3 = -29160$

Activity 3.4

a) $(1 + 0.3\char94 3)\char94 27 = 2.053$

b) $-(-2.3)\char94 3 - 4*(-23) + 12 = 116.167$

Activity 3.5

$3^3, 3^2, 3^1, 3^0, 3^{-1}, 3^{-2}, 3^{-3}, \ldots$

Activity 3.6

a) i) $a^9 b^3$ ii) $3x^5 b^{-1}$

b) i) $3ab^{-4} = \dfrac{3a}{b^4}$ ii) $a^{-4} b^{-1} = \dfrac{1}{a^4 b^1}$ iii) $4a^{-4} b^2 c^{-2} = \dfrac{4b^2}{a^4 c^2}$

Activity 3.7

a) $7^5 7^9 = 7^{5+9} = 7^{14}$

b) $(-6)^9 \bullet (-6)^{-5} = (-6)^{9+(-5)} = (-6)^4$

c) $x^{-7} \bullet x^4 = x^{-7+3} = x^{-3} = \dfrac{1}{x^3}$

<u>Activity 3.8</u>

a) $(x^4)^5 = x^{20}$

b) $(y^{-4})^2 = y^{-8} = \dfrac{1}{y^8}$

<u>Activity 3.9</u>

$$\frac{b^5}{b^3} = \frac{b \bullet b \bullet b \bullet b \bullet b}{b \bullet b \bullet b} = b^2 \qquad\qquad \frac{b^3}{b^5} = \frac{b \bullet b \bullet b}{b \bullet b \bullet b \bullet b \bullet b} = \frac{1}{b^2} = b^{-2}$$

These two examples suggest that $\qquad \dfrac{b^m}{b^n} = b^{m-n}$

<u>Activity 3.10</u>

a) $(2xy)^3 = 2^3 x^3 y^3 = 8x^3 y^3$

b) $(3x)^{-2} = 3^{-2} x^{-2} = \dfrac{1}{3^2 x^2} = \dfrac{1}{9x^2}$

<u>Activity 3.11</u>

$$\left(\frac{a}{b}\right)^3 = \frac{a}{b} \bullet \frac{a}{b} \bullet \frac{a}{b} = \frac{a^3}{b^3}$$

$$\left(\frac{a}{b}\right)^n = \underbrace{\frac{a}{b} \bullet \frac{a}{b} \bullet \frac{a}{b} \ldots \bullet \frac{a}{b}}_{n \text{ times}} = \frac{a^n}{b^n}$$

<u>Activity 3.12</u>

a) i) $10^3 = 1{,}000$ ii) $10^4 = 10{,}000$ iii) $10^5 = 100{,}000$

b) 101 digits →1 digit for the 1 and 100 digits for the one hundred zeroes.

c) $n + 1$

d) i) $1{,}000{,}000 = 10^6$ ii) $10{,}000{,}000{,}000 = 10^{10}$

e) $4{,}000{,}000 = 4 \times 10^6$

f) $50{,}000{,}000{,}000 = 5 \times 10^{10}$

<u>Activity 3.13</u>

a) i) $4,500,000,000 = 4.5 \times 10^9$ ii) $50,8000,000,000,000 = 5.08 \times 10^{13}$

b) Move the decimal point to the left until you get a number between 1 and 10. Count how many places you moved the decimal point and that becomes the exponent on the base of ten.

c) $2.8 \times 10^5 = 280,000$ d) $4.76 \times 10^8 = 476,000,000$

<u>Activity 3.14</u>

a) i) $0.003 = 3 \bullet 10^{-3}$ ii) $0.000102 = 1.02 \bullet 10^{-4}$

    iii) $0.00000034 = 3.4 \bullet 10^{-7}$ iv) $0.0000000002003 = 2.003 \bullet 10^{-10}$

b) i) 0.000076

    ii) 0.000000000403

<u>Activity 3.15</u>

1) $2.30 \times 10^{12} = 2,300,000,000,000$

2) $9.45 \times 10^{-5} = 0.0000945$

3) $2.30 \times 10^9 = 2,300,000,000$

4) $2.00 \times 10^{-10} = 0.000000000200$

<u>Activity 3.16</u>

End of Year	Amount in Account
1	$1000(1.1)^1 = \$1,100$
2	$1000(1.1)^2 = \$1,210$
3	$1000(1.1)^3 = \$1,331$
4	$1000(1.1)^4 = \$1464.10$
5	$1000(1.1)^5 = \$1610.51$
.	
.	
.	
n	$1000(1.1)^n$

<u>Activity 3.17</u>

a)

YEAR	POPULATION
1990	5000
2010	10000
2030	20000
2050	40000

b) for 2010, t = 1; for 1990, t = 0 ; for 1970, t = -1

c) for 2050, t = 3 so $P = 5000(2)^3 = 40000$

Activity 4.1

$V = 20000(0.8)^n$ represents a function because for each input, you get one and only one output.

Activity 4.2

Yes, y is a function of x in the given relation because each input has a unique output.

3→10	5→10	If you graphed the points given in the table, you would
4→10	7→10	get a horizontal line.
	10→10	

Activity 4.3

a)   i)   A function because each name as a unique number of letters.
EX/   Brian → 5
Sarah → 5
John → 4
Kim → 3

ii)   Not a function because each person may own more than one car.
EX/

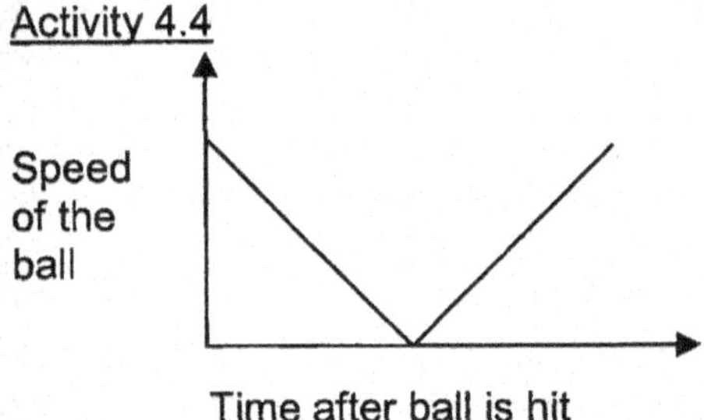

iii)   A function because each person has only one unique social security number.

iv)   A function because each set will have a unique number of elements.

b)   i)   Table: 11, 14
Rule: words: add 4 to each input value
formula: y = x + 4

ii)   Table: 19, 24
Rule: words: add 9 to each input value
formula: y = x + 9

c)   rule: x(input) ⟶ -2 ⟶ y(output)

formula: y = x - 2

Activity 4.4

Speed
of the
ball

Time after ball is hit

Activity 4.5

a) Possible set of points

x	y
0	2
1	4
2	6
3	8

b) Possible set of points

x	y
0	2
1	4
2	7
3	10

c) Possible set of points

x	y
0	2
1	5
2	7
3	8

Activity 4.6

a)   i) D(9) = 450   ii)   D(1) = 50

b)   D(10) = 500

Activity 4.7

a) $f(x) = x + 9$
        $f(3) = 3 + 9 = 12$
        $f(5) = 5 + 9 = 14$

b) $f(x) = x^2 - 4x + 7$
        $f(3) = 3^2 - 4(3) + 7 = 9 - 12 + 7 = -3 + 7 = 4$
        $f(5) = 5^2 - 4(5) + 7 = 25 - 20 + 7 = 5 + 7 = 12$

Activity 4.8

$A(0) = \$17{,}218$ is the debt for a student who graduated in the year 2000.

Activity 4.9

To graph the function on Excel, first input -3 in cell A1, -2 in cell A2, -1 in cell A3, 0 in cell A4, 1 in cell A5, 2 in cell A6, 3 in cell A7, and 4 in cell A8. In the next column, start by inputting the function,
  =4*A1^5 – 5*A1^4 – 40*A1^3
then hit enter to get the result. Fill down to have the rest of the outputs calculated. Follow the instructions in section 4-4 to graph the resulting data.

Activity 4.10

CPI   26.7 in 1953   so $\dfrac{26.7}{215.3} = 0.12$

       215.3 in 2008

So if an item costs \$1.00 in 2008, it would have cost \$0.12 in 1953.

Activity 4.11

The function is increasing, slowly at first and then rapidly, starting around 1980.

Activity 4.12

Year	CPI	Percentage Change
2002	179.88	
2003	183.96	2.30%
2004	188.9	2.70%
2005	195.3	3.40%
2006	201.6	3.20%
2007	207.34	2.80%
2008	215.3	3.80%

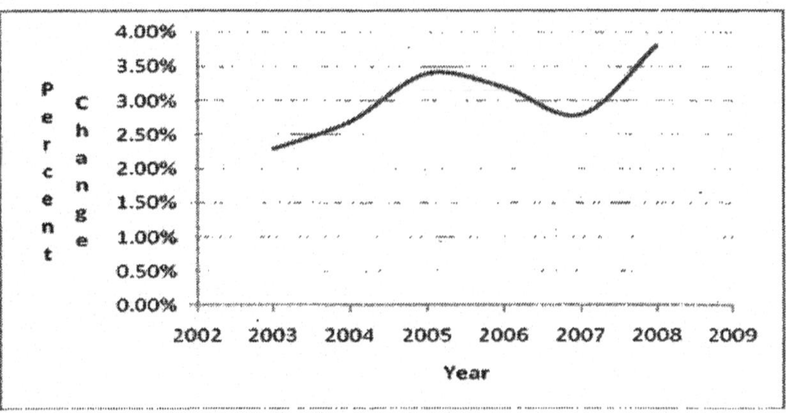

Activity 5.1

a) $5-(x+3)=1$

$5-x-3=1$

$2-x=1$

$-x=-1$

$x=1$

b) $2(x+4)=4x$

$2x+8=4x$

$8=2x$

$4=x$

c) $3x-7=4-3(x+4)$

$3x-7=4-3x-12$

$3x-7=-3x-8$

$6x=-1$

$x=-\dfrac{1}{6}$

Activity 5.2

$C(n)=\dfrac{3000}{6+n}$

$60=\dfrac{3000}{6+n}$

$60(6+n)=3000$

$360+60n=3000$

$60n=3640$

$n=44$

There are 44 additional students needed to bring the cost down to \$60/student.

Activity 5.3

a) $t(r)=\dfrac{75}{r}$

$t(40)=\dfrac{75}{40}$

$t(40)=1.875$

It will take $1\dfrac{7}{8}$ hours to get to Springfield.

b) $r(t)=\dfrac{75}{t}$

$r(1.5)=\dfrac{75}{1.5}$

$r(1.5)=50$

The average speed was 50 mph.

Activity 5. 4

Approximate Solutions:

$x=1$

$x=3.1$

$x=6.2$

$x=9.5$

Activity 5.5

Using Goal Seek:
        In A2 type, $= 3000*((1.04^{A3}-1)/(1.04-1))$
        In A3 type a guess for the value of n

        In the Goal Seek box type:
                Set Cell: A2
                To value: 50000
                By changing cell: A3

        Hit ok and get $n \approx 13$ years.

Activity 5.6

Using Solver:
        In A1, type in a guess for x
        In A2, type =2*A1^3 – 3*A1 + 1
        Open Solver and in
                • "Set Target Cell", type $A$2
                • "Value of", type 5
                • "By Changing Cells", type $A$1
        Hit Solve and find the solution to x in cell A1
        So, x = 1.65

Activity 5.7

Using Solver:
        In A1, type 47
        In A2, type $= 1000*1.035^{A1}$
        Open Solver and in
                • "Set Target Cell", type $A$2
                • "Value of", type 5000
                • "By Changing Cells", type $A$1
        Hit solve and get $n = 46.78405$.
        So it will take approximately 46.8 years to have $1000 grow to $5000.

Activity 6.1

x	y	First difference
0	2	
2	8	$8 - 2 = 6$
4	14	$14 - 8 = 6$
6	20	$20 - 14 = 6$

The data represent a linear function because the first differences are the same.

Activity 6.2

d (days in store)	c (cost in dollars)
0	$10.00
1	$9.50
2	$9.00
3	$8.50
4	$8.00
5	$7.50
6	$7.00
7	$6.50

The cost of the strawberries is a function of the number of days they sit in the store and is given by the formula: $c = 10 - 0.5d$.

The cost of the strawberries if they sat in the store for 7 days is $6.50.

Activity 6.3

The slope is -2.5 or $\dfrac{-2.5}{1}$ which means every time the demand **increases** by 1 the price **decreases** by 2.5.

Activity 6.4

Since the difference is 25, substitute that into the given equation which gives:

$$\text{wins} = 0.1059 \bullet 25 + 80.833 = 83.4805$$

Therefore, there would be 83 wins.

Activity 7.1

$$\frac{38500}{45000} = 0.86$$, so about a 14% decrease.

Activity 7.2

a) 5 acres(A) for every 2 days (d) , means $\frac{5}{2}$ acres for every 1 day: $A = \frac{5}{2}d$

b) \$90 (C) per 1 week (w): $C = 90w$

c) 72 beats (B) per 1 minute (m): $B = 72m$

Activity 7.3

a) Use 3 cups of white paint for 2 cups of blue paint to make 5 cups of paints.

b) Use 6 cups of white paint for 4 cups of blue paint to make 10 cups of paints.

c) 3 cups of white paint for every 2 cups of blue paint means $\frac{3}{2}$ cups of white paint for every 1 cup of blue paint, so

$W = \frac{3}{2}B$, substituting in the fact that we have 1 cup of white paint gives $1 = \frac{3}{2}B$ and solving for B we

get $\frac{2}{3} = B$. Now we can fill in the blanks:

Use 1 cup of white paint for $\frac{2}{3}$ cups of blue paint to make $1\frac{2}{3}$ cups of paints.

d) Now we know we have 2 total cups of paint so we will need to write one equation: white + blue = 2.
Since $W = \frac{3}{2}B$, we can substitute $\frac{3}{2}B$ for white and get $\frac{3}{2}B + B = 2$. Solving for B we get

$\frac{5}{2}B = 2$

$B = 2 \bullet \frac{2}{5} = \frac{4}{5}$

Substituting that back into $W = \frac{3}{2}B$, we get $W = \frac{3}{2} \bullet \frac{4}{5} = \frac{12}{10} = 1\frac{1}{5}$.

Now we can fill in the blanks:

Use $1\frac{1}{5}$ cups of white paint for $\frac{4}{5}$ cups of blue paint to make 2 cups of paints.

Activity 7.4

a) $F = 12I$

$$feet = \frac{12\ inches}{1\ foot} \cdot inches$$

The units do not match so the formula given is not correct.

$$feet \neq \frac{12\ inches^2}{1\ foot}$$

b) Let P = pay and h=# of hours, then $P(h) = 9h$.

The constant of proportionality is 9.

Activity 7.5

$$\frac{400\ sq\ ft}{4\ hours} = \frac{x\ sq\ ft}{6\ hours}$$

a) Set up the proportion and solve for x: $\quad 4x = 400(6)$

$$x = 600$$

Therefore, I can paint 600 square feet in 6 hours.

$$\frac{1\ inch}{15\ miles} = \frac{3.5\ inches}{x\ miles}$$

b) Set up the proportion and solve for x: $\quad 1x = 15(3.5)$

$$x = 52.5$$

Therefore, the actual distance is 52.5 miles.

Activity 7.6

To determine if L and g are directly proportional, determine if $\frac{g}{L}$ is constant. Since

$$\frac{2}{7.5708} = \frac{5}{18.9270} = \frac{7}{26.4978} = \frac{12}{45.4248} = \frac{15}{56.7810} = 0.2641728747,\ \text{then L and g are}$$

directly proportional.

Activity 7.7

If $y \alpha\, x^3$, then $y = kx^3$ and $k = \dfrac{y}{x^3}$ should be constant.

Since $\dfrac{4}{1^3} = \dfrac{256}{4^3} = \dfrac{864}{6^3} = 4$, then $y \alpha\, x^3$.

Activity 7.8

$y \,\alpha\, x$ if $\dfrac{y}{x} = k$.   Checking the values in the table, we have: $\dfrac{2}{1} = 2\,; \dfrac{8}{2} = 4\,; \dfrac{18}{3} = 6\,; \dfrac{32}{4} = 8\,; \dfrac{50}{5} = 10$

The conclusion that $y \,\alpha\, x$ is not correct because k is not constant.

Activity 7.9

We know that y cannot vary directly with x because if $y \,\alpha\, x$ then the graph would be a straight line that passes through the origin.

Activity 7.10

a) If y is directly proportional to x, then $y = kx$ and $k = \dfrac{y}{x}$ should be constant, or at least close to constant, for all ordered pairs x and y.

$$\frac{2}{5} = \frac{4}{10} = \frac{6}{15} = \frac{8}{20} = \frac{10}{25} = 0.4$$

Hence, x and y are directly proportional and the constant of proportionality is 0.4.

b) If x and y vary inversely, then $y = \dfrac{k}{x}$ and $k = xy$ should be constant.

$$24(5) = 12(10) = 8(15) = 6(20) = 25(4.8) = 120$$

Hence, x and y vary inversely and the constant of proportionality is 120.

Activity 7.11

a) $A = kB^2$ b) $V = ke^3$

c) $y = \dfrac{k}{x^3}$ d) $C = \dfrac{k}{D}$

<u>Activity 7.12</u>

a) If $y = \dfrac{k}{x^2}$, then $k = x^2 y$. If $x_1 = 5,\ y_1 = 4,\ x_2 = 10,\ y_2 = ?$, then

$$x_1^{\,2} y_1 = x_2^{\,2} y_2$$
$$5^2(4) = 10^2 y_2$$
$$100 = 100 y_2$$
$$1 = y_2$$

Therefore, when y = 1, x = 10.

b) Since the cost, C, of producing x calculators varies directly as x, we can write $C = kx$.

Next, substitute values for C and x and solve for k.

$$4050 = k(75)$$
$$54 = k$$

Now, write the equation of variation.

$$C = 54x$$

Finally, substitute 250 for x and solve for C.

$$C = 54(250)$$
$$C = 13500$$

Therefore, it would cost $13,500 to produce 250 calculators.

Activity 8.1

a) i) F = 5280M, where F is the number of feet and M is the number of miles
   ii) F = 3Y, where F is the number of feet and Y is the number of yards

b) Since both equations in i) and ii) are equal to F, we can write: 3Y = 5280M. Then solving for Y gives us: Y = 1760M.

Activity 8.2

$$\frac{\$12}{20\ lbs} \times \frac{1\ lb}{16\ oz} \times \frac{10\ oz}{1\ portion} = \$0.375\,/\,portion$$

Activity 8.3

$$54\,ft^3 \times \frac{1\ yd^3}{27\ ft^3} = 2\ yd^3$$

Activity 8.4

Unit	Symbol	Relationship to Base Unit
kilometer	km	1000 meters
hectometer	hm	100 meters
dekameter	dam	10 meters
meter	**m**	**base unit**
decimeter	dm	0.1 meter
centimeter	cm	0.01 meter
millimeter	mm	0.001 meter

Activity 8.5

a) mm or cm   b) cm or m   c) m or km   d) mm

Activity 8.6

a) $125\ mm \times \dfrac{1\ cm}{10\ mm} = 12.5\ cm$ b) $48\ cm \times \dfrac{1\ m}{100\ cm} \times \dfrac{1\ km}{1000\ m} = 0.00048\ km$

Activity 8.7

a) 6.14 mm = 0.00614 m   *move the decimal point 3 places to the left*

b) 0.75 km = 750 m   *move the decimal point 3 places to the right*

Activity 8.8

a) gm

b) kg

Activity 8.9

a) 120 g = 0.12 kg   *move the decimal point 3 places to the left*

b) 0.72 cg = 7.2 mg   *move the decimal point 1 place to the right*

Activity 9.1

a)
$$\frac{A'B'}{AB} = \frac{B'C'}{BC}$$

$A'B'BC = ABB'C'$ *cross multiply*

$$\frac{BC}{AB} = \frac{B'C'}{A'B'}$$ *divide both sides of the equation by $A'B'$ and AB*

b)
$$\frac{A'B'}{AB} = \frac{A'C'}{AC}$$

$A'B'AC = ABA'C'$ *cross multiply*

$$\frac{AC}{AB} = \frac{A'C'}{A'B'}$$ *divide both sides of the equation by $A'B'$ and AB*

c)
$$\frac{B'C'}{BC} = \frac{A'C'}{AC}$$

$B'C'AC = A'C'BC$ *cross multiply*

$$\frac{AC}{BC} = \frac{A'C'}{B'C'}$$ *divide both sides of the equation by $B'C'$ and BC*

Activity 9.2

a) a   b) a   c) c   d) adjacent

Activity 9.3

a) The ratio that includes the side opposite of A to the adjacent side of A is the tangent, so $\tan 20° = 0.364$ gives the required ratio.

b) If $\angle A = 20°$, then $\angle B = 70°$. The ratio that includes the side opposite of B to the hypotenuse is the sine, so $\sin 70° = 0.940$ gives us the required ratio.

Activity 9.4

a) 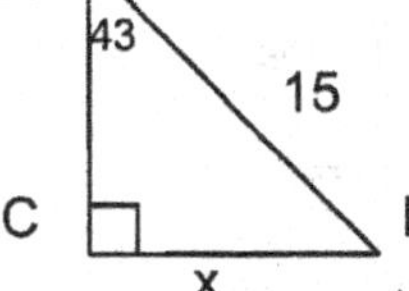

$$\sin 43° = \frac{x}{15}$$
$$15\sin 43° = x$$
$$10.2 = x$$

So the length of side a is 10.2.

---

b) 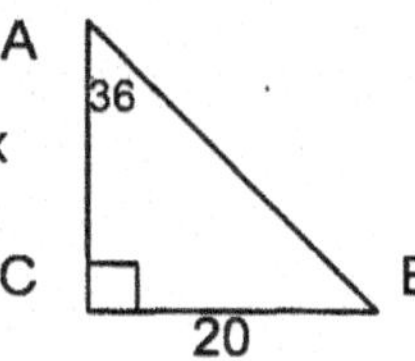

$$\tan 36° = \frac{20}{x}$$

$$x \tan 36° = 20$$

So the length of side b is 27.5.

$$x = \frac{20}{\tan 36°} = 27.5$$

Activity 9.5 Use the Pythagorean Theorem to find the third side of each triangle.

a) 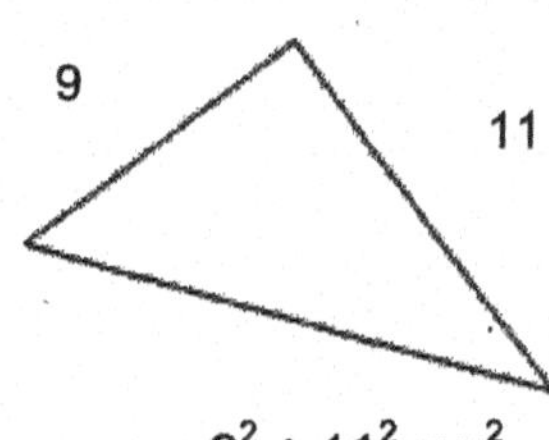

b) 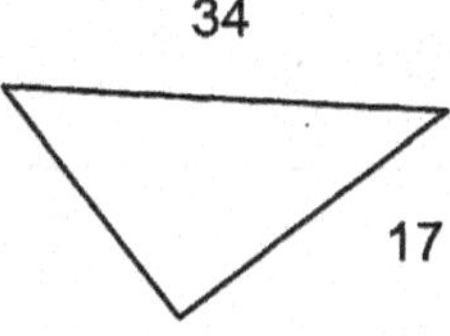

$$9^2 + 11^2 = x^2$$
$$81 + 121 = x^2$$
$$202 = x^2$$
$$14.2 \approx x$$

$$x^2 + 17^2 = 34^2$$
$$x^2 + 289 = 1156$$
$$x^2 = 867$$
$$x \approx 29.4$$

Activity 9.6

a) i) $\sin B = \dfrac{12}{15}$ so $B = \sin^{-1}\left(\dfrac{12}{15}\right) = 53.1°$

ii)
$$15^2 = a^2 + 12^2$$
$$225 = a^2 + 144$$
$$a^2 = 81$$
$$a = 9$$

b) $\sin A = \dfrac{2}{3}$ so $A = \sin^{-1}\left(\dfrac{2}{3}\right) = 41.8°$

Activity 9.7

a) To change $45.76°$ to DMS, first change .76 degrees to minutes.

$$0.76 \text{ deg} \times \frac{60 \text{ min}}{1 \text{ deg}} = 45.6 \text{ min}$$

Next, change .6 minutes to seconds.

$$0.6 \text{ min} \times \frac{60 \text{ sec}}{1 \text{ min}} = 36 \text{ sec}$$

Finally, write in DMS form: $45°45'36''$

---

b)   To change $12°25'30''$ to DD, first change 30 seconds to minutes.

$$30 \text{ sec} \times \frac{1 \text{ min}}{60 \text{ sec}} = .5 \text{ min}$$

Next, add 25 and .5 to get 25.5 minutes and then change that to degrees.

$$25.5 \text{ min} \times \frac{1 \text{ deg}}{60 \text{ min}} = .425 \text{ deg}$$

Finally, add 12 to .425 and write in DD form: $12.425°$

Activity 9.8

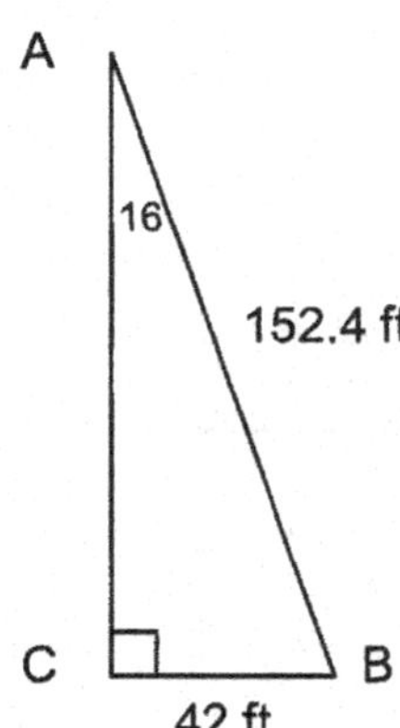

$$\angle B = 90 - 16 = 74°$$

$$\sin 16° = \frac{42}{c}$$
$$c \sin 16° = 42$$
$$c = \frac{42}{\sin 16°} = 152.4 \; ft$$

$$\tan 16° = \frac{42}{b}$$
$$b \tan 16° = 42$$
$$b = \frac{42}{\tan 16°} = 146.5 \; ft$$

Activity 9.9

a) Find the distance between the foot of the ladder and the building.

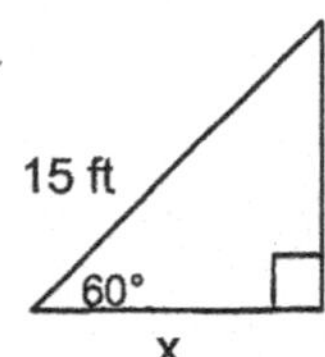

$$\cos 60° = \frac{x}{15}$$
$$x = 15 \cos 60° = 7.5 \; ft$$

b)   You can find the third side of the triangle by either using the sine of 60 degrees, or using the Pythagorean Theorem.

$$\sin 60° = \frac{x}{15}$$
$$x = 15 \sin 60°$$
$$x = 12.99 \; ft$$

$$15^2 = 7.5^2 + x^2$$
$$225 = 56.25 + x^2$$
$$168.75 = x^2$$
$$x = 12.99 \; ft$$

<u>Activity 9.10</u>

In reference to the angle of depression, we know the side adjacent (length of the shadow) and the side opposite (length of the flagpole), which means we need to use the tangent to find the measure of the angle.

$$\tan \theta = \frac{43}{75} = 0.573$$

This means that $\theta = \tan^{-1}\left(\dfrac{43}{75}\right) = 29.827 = 29°50'$ to the nearest 10 minutes.

<u>Activity 9.11</u>

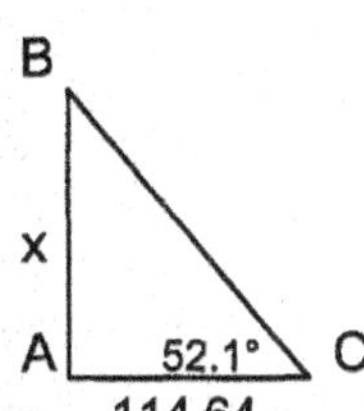

$$\tan 52.1° = \frac{x}{114.64}$$

$$114.64 \tan 52.1° = x$$

$$x = 147.26$$

Therefore, the distance between points A and B is 147.26 ft.

<u>Activity 9.12</u>

a)   Relationships: (1) $\tan 30° = \dfrac{h}{5280 + d}$ (2) $\tan 35° = \dfrac{h}{d}$

b)   Common unknowns: h Solving each equation above for h gives us,

(3) $h = (5280 + d)\tan 30°$ (4) $h = d \tan 35°$

c)   Since both equation (3) and (4) equal h, we can set the right sides of the equations equal to one another.

$$(5280 + d)\tan 30° = d \tan 35°$$

d)   Now solve the above equation for d.

$$(5280 + d)\tan 30° = d \tan 35°$$
$$5280 \tan 30° + d \tan 30° = d \tan 35°$$
$$5280 \tan 30° = d \tan 35° - d \tan 30°$$
$$5280 \tan 30° = d(\tan 35° - \tan 30°)$$
$$\frac{5280 \tan 30°}{(\tan 35° - \tan 30°)} = \frac{d(\tan 35° - \tan 30°)}{(\tan 35° - \tan 30°)}$$
$$24812.6 = d$$

To find h, substitute the value for d into equation (2).

$$\tan 35° = \frac{h}{24812.6}$$

$$h = 24812.6 \tan 35° = 17374$$

Therefore, the mountain is 17,374 feet above the road.

Activity 10.1

It looks like the solution is about x = 12 and y = 1. Your answers may vary because we can only approximate the answer.

Activity 10.2

The solution is c, (-17, 5) since both x = -17 and y = 5 satisfy both equations.

Activity 10.3

Solving the second equation, $x - 3y = -7$, for x, we get $x = -7 + 3y$. Substituting this value for x in the first equation, $3x - 2y = 0$, results in $3(-7 + 3y) - 2y = 0$.

Now solve for y.

$$3(-7 + 3y) - 2y = 0$$
$$-21 + 9y - 2y = 0$$
$$7y = 21$$
$$y = 3$$

Next, substitute y = 3 into one of the original equations.

$$x - 3y = -7$$
$$x - 3(3) = -7$$
$$x - 9 = -7$$
$$x = 2$$

Therefore, the solution is x = 2 and y = 3.

Activity 10.4

First get one of the variables to eliminate when the equations are added. To do this, multiply the first equation, $2x + 4y = 7$, by 3 and the second equation, $-3x + 3y = 2$, by –4.

$$6x + 12y = 21$$
$$12x - 12y = -8$$

Next, add these equations.

$$18x + 0 = 13$$

Finally, solve for x.

$$x = \frac{13}{18}$$

Activity 10.5

To graph $-6a - 10b = -12$, find the horizontal and vertical intercepts. When $a = 0$, $b = \dfrac{6}{5}$ and
when $b = 0$, $a = 2$. Since the intercepts are the same as the first equation, $3a + 5b = 6$, then the two graphs
will be the same. This happened because we graphed a multiple equation of the given equation.

Activity 10.6

a) $\quad \begin{vmatrix} 2 & 6 \\ -1 & 5 \end{vmatrix} = 2(5) - (-1)(6) = 10 + 6 = 16$

b) $\quad \begin{vmatrix} -1 & 1 \\ 5 & 2 \end{vmatrix} = -1(2) - 1(5) = -2 - 5 = -7$

Activity 10.7

To solve the system of equations $2x - 7y = 1$ and $-3x - y = 10$ use the formula for Cramer's Rule with a = 2,
b = -7, c = 1, d = -3, e = -1 and f = 10.

$$x = \dfrac{\begin{vmatrix} 1 & -7 \\ 10 & -1 \end{vmatrix}}{\begin{vmatrix} 2 & -7 \\ -3 & -1 \end{vmatrix}} = \dfrac{1(-1) - (-7)(10)}{2(-1) - (-3)(-7)} = \dfrac{-1 - (-70)}{-2 - 21} = \dfrac{69}{-23} = -\dfrac{69}{23} = -3$$

$$y = \dfrac{\begin{vmatrix} 2 & 1 \\ -3 & 10 \end{vmatrix}}{23} = \dfrac{2(10) - 1(-3)}{-23} = \dfrac{20 + 3}{-23} = -1$$

Therefore, the solution is x = -1 and y = -3.

Activity 10.8

To solve the system of equations $y = 5 - 3x$ and $9x + 3y = 15$, first put the first equation in
Ax + By = C form, $3x + y = 5$.

Next, multiply the first equation by -3 and add the two equations.

$$-9x - 3y = -15$$
$$9x + 3y = 15$$
$$0 + 0 = 0$$

Since the result is 0 = 0 then the system has an infinite number of solutions.

<u>Activity 10.9</u>

a) The new point of intersection is approximately (200, 120) or (200, $1.20).

b)
$$P = -0.5Q + 230$$
$$P = 0.5Q + 20$$

Since both expressions equal P, we can set the two expressions equal to one another and solve.

$$-0.5Q + 230 = 0.5Q + 20$$
$$210 = 1Q$$

Substituting 210 for Q, gives us

$$P = 0.5(210) + 20 = 125$$

Verifying the solution (210, 125) by using the other equation we get:

$$125 = -0.5(210) + 230$$
$$125 = 125$$

Therefore, the algebraic solution is (210, 125) or (210, $1.25).

<u>Activity 10.10</u>

a) Since the supply curve translates 60 units to the right, we will subtract 60 from the quantity.

$$P = 0.5(Q - 60) + 20$$
$$P = 0.5Q - 30 + 20$$
$$P = 0.5Q - 10$$

b) Verifying the new point of intersection (190, 85), we get:

$$P = 0.5Q - 10$$
$$85 = 0.5(190) - 10$$
$$85 = 85$$

<u>Activity 10.11</u>

Answers will vary, but here is one example. You could say that any change that shifts the supply curve inward to the left and does not affect the demand curve, will raise the equilibrium price and decrease the equilibrium quantity.

Activity 11.1

Choices b and c are quadratic functions because they can be put in the form $y = ax^2 + bx + c$.

Activity 11.2

a)

n	T	1st difference	2nd difference
0	0		
1	0	0	
2	1	1	1
3	3	2	1
4	6	3	1
5	10	4	1
6	15	5	1
7	21	6	1

b) The function is quadratic because the second differences are the same. Also, if we distribute the n in

the formula $t(n) = \dfrac{(n-1)n}{2}$, we get $t(n) = \dfrac{n^2 - n}{2} = \dfrac{1}{2}n^2 - \dfrac{1}{2}n$. Since $t(n)$ is in the form

$f(x) = ax^2 + bx + c$, it is a quadratic function.

Activity 11.3

a) <u>Cooling Coffee</u>: This table does not represent a linear or a quadratic function.

b) <u>Cooking Times for Turkey</u> : This table represents a linear function because the 1st difference is constant.

c) <u>Baby Before Birth</u>: This table does not represent a linear or a quadratic function

---

Activity 11.4

a) First, set the equation equal to 0: $\qquad 4x^2 + 5x - 1 = 0$

Second, find the discriminant: $\quad b^2 - 4ac = 5^2 - 4(4)(-1) = 41$ Since the discriminant is positive, $4x^2 + 5x - 1 = 0$ has two solutions.

Third, use the quadratic formula to solve the equation.
(a = 4, b = 5, c = -1, and discriminant = 41)

$$x = \frac{-5 \pm \sqrt{41}}{2(4)} = \frac{-5 \pm \sqrt{41}}{8}$$

$$x = \frac{-5 + \sqrt{41}}{8} = 0.175 \quad and \quad x = \frac{-5 - \sqrt{41}}{8} = -1.43$$

<u>Activity 11.4 continued</u>

b)   First, set the equation equal to 0: $4x^2 - 4x + 1 = 0$

Second, find the discriminant: $b^2 - 4ac = (-4)^2 - 4(4)(1) = 0$ Since the discriminant is zero, $4x^2 - 4x + 1 = 0$ has one solution.

Third, use the quadratic formula to solve the equation.
(a = 4, b = -4, c = 1, and discriminant = 0)

$$x = \frac{4 \pm \sqrt{0}}{2(4)} = \frac{4}{8} = 0.5$$

<u>Activity 11.5</u>

a)   i)   opens down since $a < 0$ ii)   (0, -1)

iii) $x = \frac{-b}{2a} = \frac{3}{2(-2)} = \frac{3}{-4} = -0.75$ line of symmetry: x = -0.75

iv) $y = -2x^2 - 3x - 1 = -2(-0.75)^2 - 3(-0.75) - 1 = 0.125$ vertex: (-0.75, 0.125)

b)   First, find the value for p, using the formula for the line of symmetry : p = $\dfrac{-b}{2a}$.

$$p = \frac{-b}{2a} = \frac{-260.3}{2(-320.5)} = 0.406$$

Next, substitute the value for p into the given revenue equation.

R = -320.5(0.406)^2 + 260.3(0.406) = 52.85

Therefore, you should charge $0.406 to get maximum revenue of $52.85.

<u>Activity 11.6</u>

Refresh your memory on how to use Goal Seek and then follow these steps on Excel.

In cell A1, type: =A2*A2+4*A2-9

In cell A2, input a guess for the solution: 5   (put whatever number you want)

Open Goal Seek and type the following:
     Set cell: A1
     To value: 10
     By changing cell: A2

Hit OK and the <u>solution of 2.795886</u> should appear in cell A2.

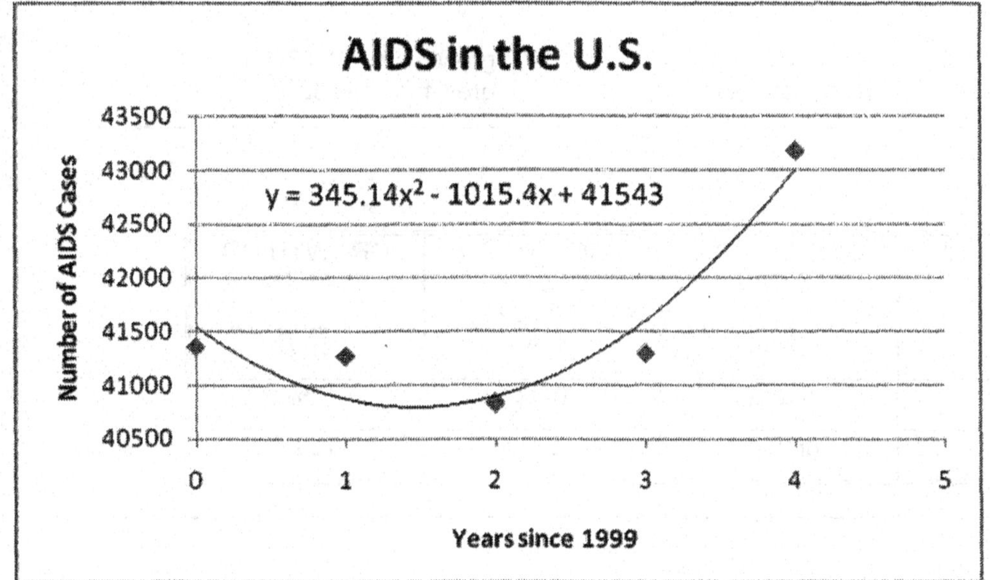

AIDS in the U.S.
Number of AIDS Cases
43500
43000
42500
42000
41500
41000
40500
y = 345.14x^2 - 1015.4x + 41543
0
1
2
3
4
5
Years since 1999

Activity 12.1

a)   growth since 1.98 > 1   b)   decay since 0.98 < 1   c)   growth since 1.33 > 1
     growth rate is 98%   decay rate is 2%   growth rate is 33%

Activity 12.2

FUNCTION	INITIAL VALUE	GROWTH OR DECAY	GROWTH OR DECAY FACTOR	GROWTH OR DECAY RATE
$B = 15(1.59)^t$	15	growth	1.59	59%
$y = 100(0.98)^x$	100	decay	0.98	2%
$y = 1.5(2.3)^x$	1.5	growth	2.3	130%

Activity 12.3

a)   The data in the table represents an exponential function because the ratios are equal.

$$\frac{7.8}{2.6} = \frac{23.4}{7.8} = \frac{70.2}{23.4} = 3$$

To write the exponential function, we need the initial value and the growth factor. We just found the growth factor by finding the common ratio, 3.

The initial value can be read off the table when x = 0, y = 2.6. Therefore, the exponential function would be

$$y(x) = 2.6(3)^x.$$

b)   The data in the table does not represent an exponential function because the ratios are not constant.

$$\frac{9}{5} = 1.8 \qquad \frac{21}{9} = 2.33 \qquad \frac{43}{21} = 2.04$$

<u>Activity 12.4</u>

Using Excel as shown below, gives the exponential function that best fits the data as $y = (2 \times 10^6)e^{0.0354x}$.

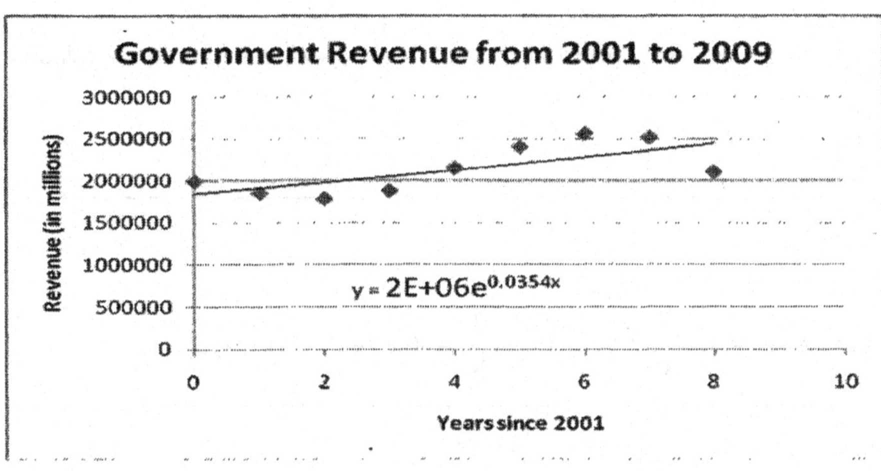

<u>Activity 12.5</u>

a) If you invested \$2,500 at an annual interest rate of 5%, it would take $\dfrac{70}{5} = 14$ years to double your

$2,500. *(Notice you do <u>not</u> change the percent to a decimal to determine the doubling time.)*

b) If the population of the United States is growing at the rate of 1.2% each year and we continue to grow

at this rate, it would take $\dfrac{70}{1.2} = 58.3$ or $58\frac{1}{3}$ years for the population to double.

Activity 13.1

To find the solution of $\log_2(\frac{1}{4}) = y$, rewrite in exponential form: $2^y = \frac{1}{4}$. Next, ask yourself, 2 raised to what power would equal $\frac{1}{4}$. The answer is y = -2.

Activity 13.2

a) log 50 = 1.70   b) In Excel, enter =log($\frac{3}{8}$, 5), so $\log_5(\frac{3}{8})$ = -0.60942   b) ln 35 = 3.56

Activity 13.3

a) $\frac{1}{2}\log_b x - \log_b(1+x^2)$ b) $2\log_b x - 3\log_b y$

$\log_b x^{\frac{1}{2}} - \log_b(1+x^2)$ Pr*operty* 3 $\log_b x^2 - \log_b y^3$ Pr*operty* 3

$\log_b\left(\dfrac{x^{\frac{1}{2}}}{(1+x^2)}\right)$ Pr*operty* 2 $\log_b\left(\dfrac{x^2}{y^3}\right)$ Pr*operty* 2

Activity 13.4

a) $4^x = 105.8$

$\log 4^x = \log 105.8$ Pr*operty* 4

$x\log 4 = \log 105.8$ Pr*operty* 3

$x = \dfrac{\log 105.8}{\log 4}$ *divide both sides by* $\log 4$

$x \approx 3.363$

b) $254.7(1.025)^x = 290$

$1.025^x = 1.139$ *divide both sides by* 254.7 *to isolate the* var*iable*

$\log 1.025^x = \log 1.139$ Pr*operty* 4

$x\log 1.025 = \log 1.139$ Pr*operty* 3

$x = \dfrac{\log 1.139}{\log 1.025}$ *divide both sides by* $\log 1.025$

$x \approx 5.271$

---

a) $e^x = 7.2$

$\ln e^x = \ln 7.2 \qquad Pr operty\ 4$

$x \ln e = \ln 7.2 \qquad Pr operty\ 3$

$x = \ln 7.2 \qquad \ln e = 1$

$x \approx 1.97$

b) $45 = 20e^{0.025t}$

$2.25 = e^{0.025t} \qquad\qquad divide\ both\ sides\ by\ 20$

$\ln 2.25 = \ln e^{0.025t} \qquad Pr operty\ 4$

$\ln 2.25 = 0.025t(\ln e) \qquad Pr operty\ 3$

$\ln 2.25 = 0.025t \qquad \ln e = 1$

$t = \dfrac{\ln 2.25}{0.025} \qquad\qquad divide\ both\ sides\ by\ 0.025$

$t \approx 32.4$

Activity 13.6

To solve $3^x = 7$ for x, locate 7 on the y-axis, then draw a horizontal line to determine where 7 hits the curve, and then drop a vertical line down to find the x value. So $x \approx 1.75$.

Activity 13.7

$1.026 = e^x$

$\ln 1.026 = \ln e^x \qquad Pr operty\ 4$

$\ln 1.026 = x \ln e \qquad Pr operty\ 3$

$\ln 1.026 = x \qquad \ln e = 1$

$x \approx 0.0257$

So, $P(t) = 67.38(e^{0.0257})^t = 67.38e^{0.0257t}$

Activity 14.1

a) $a_{20} = 66$ $d = 4$ b) $a_{20} = 66$ $d = 4$

      $a_{21} = 70$ $a_{19} = 66 - 4 = 62$

      $a_{22} = 74$ $a_{18} = 62 - 4 = 58$

                            $a_{17} = 58 - 4 = 54$

Activity 14.2

Pattern: General: $a_n = a_1 + d(n-1)$

$a_n = 5 + 3(n-1)$

$a_{100} = 5 + 3(100 - 1) = 5 + 3(99)$

$a_{100} = 302$

Activity 14.3

Yes, it would be an arithmetic sequence and the difference would be double the original difference.

Activity 14.4

Let s represent the second term then the sequence would be a, s, b.

To find the difference, we need to subtract a from b and divide by 2: $\dfrac{b-a}{2} = d$. Since a is the 1st term, then

the 2nd term would be $a + \dfrac{b-a}{2}$. Adding these we get, $\dfrac{2a}{2} + \dfrac{b-a}{2} = \dfrac{a+b}{2}$.

Therefore, the 2nd term is $\dfrac{a+b}{2}$.

Activity 14.5

Pattern: $a_n = 2 \bullet 3^{n-1}$ 100th term when r = 3: $a_{100} = 2 \bullet 3^{100-1} = 2 \bullet 3^{99} = 3.44 \times 10^{47}$

General 100th term: $a_{100} = a_1 \bullet r^{100-1}$

Activity 14.6

Given the sequence $3, 4, \ldots$ we need to find the common ratio, r, by determining what times 3 equals 4.

Using the equation, $4 = 3r$, we find $r = \dfrac{4}{3}$. To determine the third term, we multiply 4 by $\dfrac{4}{3}$ to get $\dfrac{16}{3}$.

Activity 14.7

Given that 1 is the first term of a geometric sequence and 4 is the third term, we can let s be the second term.

We know that s is either found by multiplying the first term by the common ratio, r, or by dividing the third term by r. This gives us the following equations: $S = 1 \bullet r$ and $S = \dfrac{4}{r}$.

Thus we have, $1 \bullet r = \dfrac{4}{r}$ and solving for r gives us $r^2 = 4$ or $r = 2$. Therefore, the second term is equal to 2.

Activity 14.8

a) $1 + 2 + 3 + 4 + \ldots + 100$ This sum is an arithmetic series with $a = 1$ and $b = 100$.

To find the number of terms, n:
$$a_n = 1 + 1(n-1)$$
$$a_n = 1 + n - 1$$
$$100 = n$$

Using the formula for the sum of an arithmetic series, we get $S = \dfrac{100(1+100)}{2} = 5050$.

b) $1 + \dfrac{1}{2} + \dfrac{1}{4} + \dfrac{1}{8}$ This sum is a geometric series with $a = 1$, $r = 0.5$, and $n = 4$.

Using the formula for the sum of a geometric series, we get $S = \dfrac{1(0.5^4 - 1)}{(0.5 - 1)} = 1.875$.

Activity 14.9

a) $10 + 11 + 12 + \ldots + 50 = \displaystyle\sum_{i=10}^{50} i$

b) $1 + \dfrac{1}{2} + \dfrac{1}{4} + \ldots + \dfrac{1}{128} = \displaystyle\sum_{n=1}^{8} \dfrac{1}{2^{n-1}}$

Activity 14.10

We have the geometric sequence 2, 6, 18, 54, 162, ..., 13122 where $a_1 = 2$ and $r = 3$. Therefore, the explicit formula is: $a_n = 2 \bullet 3^{n-1}$.

Activity 14.11

Jan. 1 →interest 1st year → $3120.00

So $a = 3120$, $r = 1.04$, and $n = 4$, we get $S = \dfrac{3120(1.04^4 - 1)}{(1.04 - 1)} = \$13248.97$.

Activity 15.1   a)   60 steps   b)   0

Activity 15.2

The solution to this activity is left to the student to complete since a protractor and ruler are required.

Activity 15.3

The magnitude of vector, **V**, is 38 lb and the direction is 40° South of West.

Activity 15.4

a)   i) 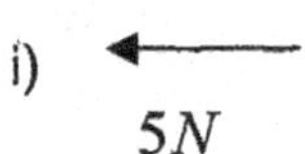

ii)
3 mph

b) 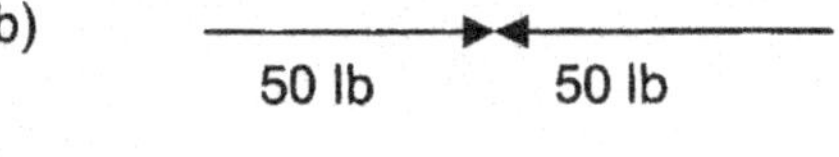

The sum is 0 lb.

c)
50 lb   50 lb

The sum is 100 lb.

d)   In vector addition 50 + 50 can take on the values of 0 through 100.

Activity 15.5

To find the resultant vector, **R**, first redraw the diagram, using the start-to-finish method.

12 lb

5 lb

R

Use the Pythagorean Theorem to find the magnitude of the resultant.

$$R^2 = 5^2 + 12^2$$
$$R = \sqrt{5^2 + 12^2} = 13$$

Therefore, the resultant vector, **R**, has a magnitude of 13 lb.

To find the direction, solve for θ, the angle made by the 12 lb vector and the resultant.

$$\tan\theta = \frac{5}{12}$$
$$\theta = \tan^{-1}\left(\frac{5}{12}\right) = 22.6°$$

Therefore, the direction is 22.6° South of East.

<u>Activity 15.6</u>

a)   2   b)   1   c)   4

<u>Activity 15.7</u>

a)   The finishing point of vector **A** is (3, 2) so **A** = 3i + 2j. The finishing point of vector **B** is (3, -1) so **B** = 3i - j.

b) $|A| = \sqrt{3^2 + 2^2} = \sqrt{13} = 3.6$ and $|B| = \sqrt{3^2 + (-1)^2} = \sqrt{10} = 3.2$

<u>Activity 15.8</u>

To write **V** in component form, find the x-component of V (V_x) and the y-component of V (V_y).

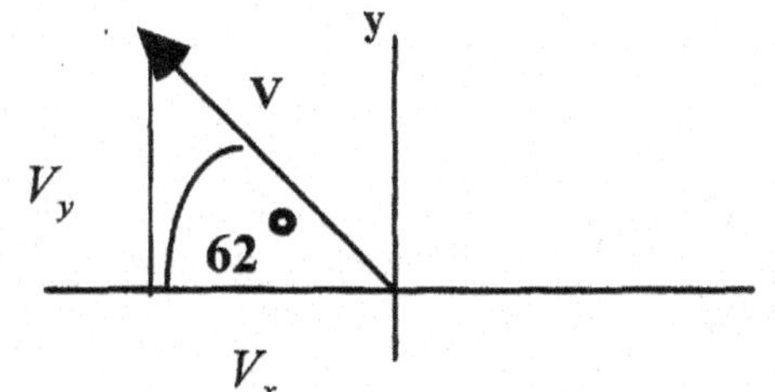

$$\cos 62° = \frac{V_x}{8.2} \qquad \sin 62° = \frac{V_y}{8.2}$$
$$8.2\cos 62° = V_x \qquad 8.2\sin 62° = V_y$$
$$3.8 = V_x \qquad 7.2 = V_y$$

Therefore, **V** = -3.8 i + 7.2 j. Notice that the x-component is negative because it is to the left of the origin.

<u>Activity 15.9</u>

R = A + B = (3.5 − 2) i + (-6 + 9) j = 1.5 i + 3 j

By the Pythagorean Theorem, the magnitude of the resultant **R** is

$$|R| = \sqrt{1.5^2 + 3^2} = 3.4$$

<u>Activity 15.10</u>

First, draw a diagram.

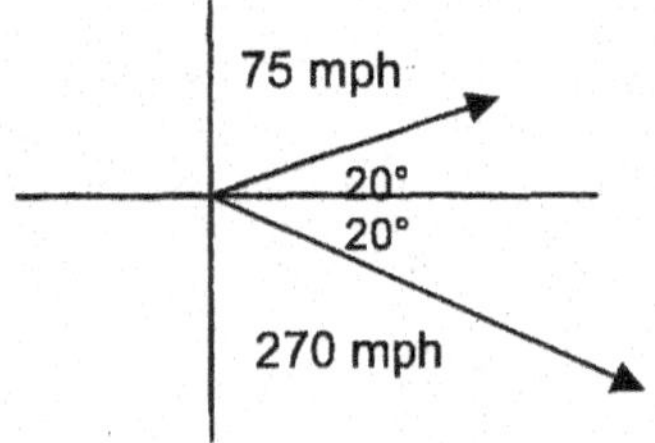

Second, write the magnitude of the airplane in component form.

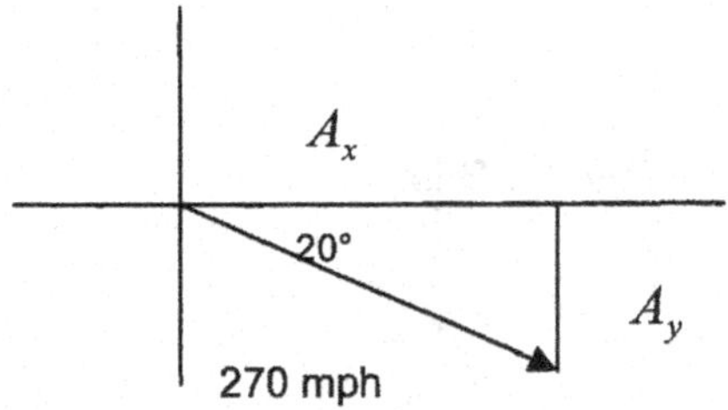

$$\cos 20° = \frac{A_x}{270} \qquad \sin 20° = \frac{A_y}{270}$$

$$270\cos 20° = A_x \qquad 270\sin 20° = A_y$$

$$253.7 = A_x \qquad 92.3 = A_y$$

$$\mathbf{A} = 253.7\,\mathbf{i} - 92.3\,\mathbf{j}$$

Third, write the magnitude of the wind in component form.

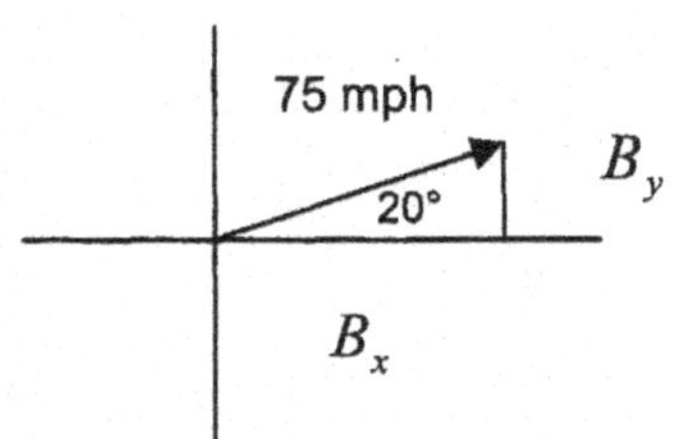

$$\cos 20° = \frac{B_x}{75} \qquad \sin 20° = \frac{B_y}{75}$$

$$75\cos 20° = B_x \qquad 75\sin 20° = B_y$$

$$70.5 = B_x \qquad 25.7 = B_y$$

$$\mathbf{B} = 70.5\,\mathbf{i} + 25.7\,\mathbf{j}$$

Fourth, add **A** and **B**.

$$\mathbf{R} = \mathbf{A} + \mathbf{B} = (253.7 + 70.5)\,\mathbf{i} + (-92.3 + 25.7)\,\mathbf{j} = 324.2\,\mathbf{i} - 66.6\,\mathbf{j}$$

Finally, write **R** in magnitude-angle form.

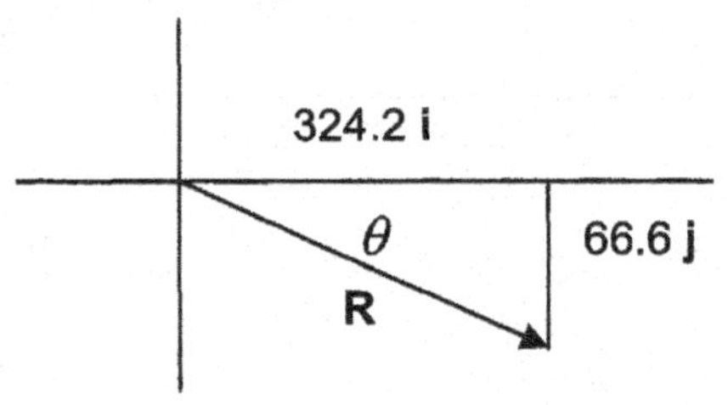

$$|R| = \sqrt{324.2^2 + (-66.6)^2} = 331$$

$$\tan\theta = \frac{66.6}{324.2}$$

$$\theta = \tan^{-1}\left(\frac{66.6}{324.2}\right) = 11.6°$$

Therefore, the airplane's speed is 331 mph and its direction is 11.6° South of East.